CHEMICAL KINETICS OF HOMOGENEOUS SYSTEMS

ROBERT SCHAAL

ENSCP – University of Paris

CHEMICAL KINETICS
OF HOMOGENEOUS
SYSTEMS

D. REIDEL PUBLISHING COMPANY

DORDRECHT-HOLLAND / BOSTON-U.S.A.

LA CINÉTIQUE CHIMIQUE HOMOGÈNE
First published by Presses Universitaires de France, Paris, 1971
Translated from the French by J. T. Edward

Library of Congress Catalog Card Number 73-94455

ISBN 90 277 0446 5

Published by D. Reidel Publishing Company,
P.O. Box 17, Dordrecht, Holland

Sold and distributed in the U.S.A., Canada, and Mexico
by D. Reidel Publishing Company, Inc.
306 Dartmouth Street, Boston,
Mass. 02116, U.S.A.

Printed in The Netherlands by D. Reidel, Dordrecht

CONTENTS

Introduction   VII

Abbreviations and Symbols   IX

1. Formal Kinetics   1
2. Kinetic Theories of Elementary Reactions   36
3. Experimental Methods   51
4. Reactions in the Gas Phase   78
5. Reactions in Solution   117
6. Homogeneous Catalytic Reactions   152

Bibliography   180

INTRODUCTION

Chemical kinetics aims to explain the factors governing the change with time of chemical systems. The results enable one to define optimum technico-economic conditions (such as the choice of batch or continuous processes; of concentration, temperature, and pressure; of whether to use a catalyst) for the preparation of products, so that kinetics is intimately bound up with many processes of chemical industry (production, explosions, combustion, propulsion in air and in space).

On another level, kinetic studies are indispensable for understanding reaction mechanisms, which implies a detailed knowledge of the different chemical intermediates (possibly very transitory) of a chemical reaction. But in practice it is rarely possible to work with microscopic quantities of reagents and, with the exception of crossed molecular beams, all methods give only statistical results concerning a large number of molecules. Because of this restriction, it has not always been possible to establish conclusively a reaction mechanism, even for reactions apparently simple.

Numerous attempts have been made to calculate rate constants from the physical properties of the participating molecules; but the introduction of the 'time' factor into calculations of the distribution of energies of chemical processes makes this very difficult, so that the elucidation of mechanisms still depends almost entirely on experimental studies. However, several theories have been elaborated which, in giving a more and more precise picture of the reaction process, have proved very fruitful, and have become indispensable in designing experiments.

This work aims to give an overall view of present tendencies in the field of chemical kinetics in homogeneous media, and to draw attention to topics capable of making an important contribution to modern chemistry. Limited in its number of pages, it is not an exhaustive treatise, and contains only the examples necessary to understand the text; a list of books covering the material in greater detail is given at the end of the text.

ABBREVIATIONS AND SYMBOLS

$\mathscr{V}$	rate
v	specific rate
ε	dielectric constant
D	diffusion coefficient
I	ionic strength
μ	dipole moment
k	specific rate coefficient
K	equilibrium constant
Log	natural logarithm to base e
log	decadic logarithm to base 10
s	second
M	mole l^{-1}
h	Planck's constant
$\mathscr{K}$	Boltzman's constant
N	Avogadro number
R	gas constant
T	absolute temperature
θ	temperature Celsius
σ	collision diameter, Hammett substituent constant
ρ	Hammett reaction constant

CHAPTER 1

FORMAL KINETICS

1.1. General

The majority of experimental studies are made in closed systems (without gain or loss of matter), isothermal (exchanging heat with the exterior), and homogeneous. However, in industrial practice, and in the study of rapid reactions and of strongly exothermic processes, one encounters kinetically 'open systems', which imply the existence of concentration gradients (continuous flow processes) or of thermally adiabatic processes (flames, explosions, etc.). The following definitions apply to closed systems, and their application to open systems will not be detailed until Chapter 3.

A. *Rate of Reaction*

The *rate* of a chemical reaction of the type

$$\sum v_i A_i = 0 \qquad A + B \rightarrow C + D$$

(with the convention that the stoichiometric coefficients v_i are negative for reactants and positive for products) is defined by the relation

$$\mathscr{V} = \frac{1}{v_i} \frac{dn_i}{dt}$$

where n_i represents the number of molecules of substance i at the time t.

The much-used term '*specific rate*' designates the quantity

$$\mathscr{V} = \frac{1}{v_i V} \frac{dn_i}{dt}$$

where V represents the volume of the reaction mixture at the time t.

Notes: (1) These definitions are general, and are to be preferred to definitions based on a reactant or a product, which are related to the reaction rate by the relation

$$\mathscr{V}_i = \pm v_i \mathscr{V}.$$

(2) For the case of the volume V remaining constant

$$\mathscr{V} \rightarrow \frac{1}{v_i} \frac{dC_i}{dt} \quad \text{with} \quad C_i = \frac{n_i}{V}.$$

(3) If the volume of the reaction mixture varies during the time of the reaction, it becomes necessary to take account of its effect (directly, and indirectly by diffusion) in changing concentration

$$dC = \left(\frac{\partial C}{\partial t}\right)_{VD} dt + \left(\frac{\partial C}{\partial t}\right)_{VR} dt + \left(\frac{\partial C}{\partial V}\right)_{RD} dV.$$

The indices V, D, R indicate the constancy of the quantity corresponding to volume, diffusion or reaction (no chemical change). Now

$$\left(\frac{\partial C_i}{\partial t}\right)_{VD} = \pm v_i \mathscr{V} \qquad \left(\frac{\partial C_i}{\partial V}\right)_{RD} = -\frac{n_i}{V^2}$$

and according to the second law of Fick

$$\left(\frac{\partial C}{\partial t}\right)_{RV} = \nabla^2 (D_i C_i)$$
$$= \frac{\partial^2}{\partial x^2} (D_i C_i) + \frac{\partial^2}{\partial y^2} (D_i C_i) + \frac{\partial^2}{\partial z^2} (D_i C_i).$$

This results in

$$\frac{dC_i}{dt} = v_i \mathscr{V} - \frac{n_i}{V^2} \frac{dV}{dt} + \nabla^2 (D_i C_i).$$

(4) The extent of a reaction with time is defined by

$$X_i = \frac{n_i - n_i^0}{v_i}$$

where n_i is the initial number of moles of i, and the fractional conversion corresponds to

$$f = \frac{X_i}{X_{i\,max}}$$

where $X_{i\,max}$ is the maximum conversion for the component i. If the component chosen is a limiting reactant 1

$$X_{max} = -\frac{n_l^0}{v_i}$$

since the reaction stops when $n_1 = 0$. (In effect, the reactants cannot be used up in stoichiometric proportions and some must still be present when the reaction ends.) The extent of reaction varies between 0 and 1 and is independent of the choice of reactant or product; it is sometimes expressed as a percentage completed.

In the case of an equilibrium reaction, $X \to X_\infty$, which is less than X_{max}. The relation $n = X/X_\infty$ is then utilized under the term *efficiency* of the reaction. (For an irreversible reaction $X_\infty = X_{max}$.)

These latter notations can be expressed with molar concentrations if the volume of the reaction mixture does not change with time.

B. *Order of Reaction and Rate Coefficient*

Experiment has shown that for some chemical reactions, the specific rate is related in simple fashion to the concentrations of reactant by

$$\mathscr{V} = k \prod_i C_i^{\alpha_i}.$$

The reaction then has an experimental order represented by a sum $\sum \alpha_i$ for the reactants, which is generally different from the sum of the stoichiometric coefficients of these reactants, and which is sometimes wrongly named 'molecularity'. The experimental coefficient α_i represents the partial order of the reaction with respect to the reactant *i*.

The experimental order generally remains constant with time, but sometimes varies with the extent of the reaction (complex reactions); it is then preferable to measure the initial order, because at zero time the products of the reaction are not able to complicate the reaction mechanism (chemical equilibria, autocatalytic reactions, etc.).

The numerical value of the order of a reaction can be integral or fractional; in homogeneous media it varies usually between 0.5 and 4, but can rise to 7 or 8 according to certain authors.

In the preceding expression, k is the specific rate coefficient having the dimensions

$$\text{time}^{-1} \text{ concentration}^{1-\Sigma\alpha_i}$$

and characterises the corresponding reaction both qualitatively and quantitatively.

This constant k is immutable for radioactive disintegrations (hence its application to the measurement of time by atomic 'clocks'), but depends essentially on temperature for chemical reactions, qualified for this reason as thermal reactions.

The variation of the rate coefficient with temperature is given by the empirical formula of Arrhenius (1889)

$$k = A\,e^{-E/RT}$$

where A is the frequency factor and E the activation energy of the reaction. Modern theoretical conceptions allow us to explain this law and indicate its limits.

Note: Homogeneous reactions which have orders varying with time can often be described mathematically, but the relations between specific rate and concentrations no longer have the simple forms shown above (complex reactions).

1.2. Simple Reactions

These are reactions which show first, second, or third order behaviour, however, they are not necessarily 'elementary' at the molecular level. A description of them is very important because of their frequent occurrence; in addition, it is often possible to treat more complex cases with the help of corresponding mathematical expressions.

A. *First Order Reactions*

All radioactive disintegrations are rigorously first-order, as are a certain number of chemical reactions (pyrolyses, isomerisations, etc.).

Schema-type: A → Products.

The rate equation can be written in two ways:

(a) $\quad \mathscr{V} = k\,|A| = -\dfrac{d\,|A|}{dt}$

of which the solution, taking account of initial conditions, is

$$kt = \text{Log}\,\frac{|A_0|}{|A|}$$

A = concentration of A at t.

A_0 = concentration of A at $t=0$

or: $|A| = |A_0|\,e^{-kt}$.

The concentration of the reactant decreases exponentially as a function of time, so that the rate decreases in the same way from the beginning to the end of the reaction.

(b) The following notation is more often used and facilitates calculations in more complicated cases.

If a is the initial concentration and x represents the concentration used up at the time t, it follows

$$\mathscr{V} = \frac{dx}{dt} = k(a - x)$$

of which the solution is

$$\text{Log}\ \frac{a}{a - x} = kt \quad \text{or} \quad x = a(1 - e^{-kt}).$$

The rate coefficient, homogeneous for time^{-1}, characterises a first-order reaction, and is preferably expressed in seconds^{-1}, but the minute or even the year may also be used as units of time, in nuclear-chemistry in particular.

The reaction period or time of half-reaction is the time necessary for the number of reacting particles (molecules, desintegrating atoms) to be reduced to half the initial value, and is given by

$$t_{1/2} = \frac{\log 2}{k} = 0.693/k.$$

The average life of a particle corresponds in fact to the time necessary for the number to diminish in the ratio of e to 1 and is calculated by the formula for the average

$$t_m = \frac{1}{C_0} \int_0^\infty C\ dt = 1/k.$$

These two parameters, both independant of the initial concentration, are both characteristic of a first-order reaction.

B. *Second Order Reactions*

Schema-type: $A + B \rightarrow$ Products.

If the initial concentrations in the reaction mixture are a and b, the rate equation is

$$\frac{dx}{dt} = k(a - x)(b - x).$$

If $b > a$, A is the limiting reactant and the reaction stops when $x = a$.

The integration of this differential equation, effected after separation of variables and decomposition into rational fractions, yields

$$\frac{dx}{(b - a)}\left(\frac{1}{a - x} - \frac{1}{b - x}\right)$$

$$= k\,dt \rightarrow \frac{1}{b - a}\,\text{Log}\,\frac{b - x}{a - x} = kt + C.$$

The constant of integration is calculated taking account of initial conditions ($t = 0 \rightarrow x = 0$), so that the general solution is

$$\frac{1}{b - a}\,\text{Log}\,\frac{a(b - x)}{b(a - x)} = kt.$$

If, on the other hand, the two reactants are present

initially in the same concentration, or if the reaction is a second-order process for the decomposition of a single reactant (type: $2A \rightarrow$ Products), the rate equation becomes

$$\frac{dx}{dt} = k(a-x)^2.$$

There is no limiting reactant in this case, and integration of this equation (taking account of initial conditions) gives

$$kt = \frac{1}{a-x} - \frac{1}{a} = \frac{x}{a(a-x)}.$$

The dimensions (time^{-1} concentration^{-1}) of the coefficient of this equation are characteristic of a second-order reaction.

The notion of period is applicable to this case only when $a = b$

$$t_{1/2} = \frac{1}{ka}.$$

In the general case, it is possible to define a period only for the extent of the reaction when $a > b$, B is the limiting reactant, $X_{max} = b$ and $f = x/b$. Since the general expression for a second-order reaction can be written in the form

$$kt = \frac{1}{b-a} \text{Log} \frac{a(1-f)}{a-fb},$$

for

$$f = \tfrac{1}{2} \rightarrow t_{1/2} - \frac{1}{k(b-a)} \text{Log} \frac{a}{2a-b}.$$

This equation is little used because it depends on initial concentrations and is not uniquely characteristic of a second-order reaction.

Note: Certain reactions, which follow experimentally the second-order rate law, are of the type $A + 2B$ or $A + 3B \rightarrow$ Products. The rate equations are then

$$dx/dt = k(a - x)(b - 2x), \quad dx/dt = k(a - x)(b - 3x)$$

leading to the solutions

$$kt = \frac{1}{2a - b} \operatorname{Log} \frac{b(a - x)}{a(b - 2x)}$$

$$kt = \frac{1}{3a - b} \operatorname{Log} \frac{b(a - x)}{a(b - 3x)}.$$

Second-order reactions are frequently encountered in both the gas phase and in solution (pyrolyses, hydrogenation of olefins, saponification of esters, etc.).

C. *Third Order Reactions*

Unlike second-order reactions, third-order reactions are rare: in the gas phase, with the exception of recombinations of free atoms $(2H^{\bullet} + H_2 \rightarrow 2H_2)$ (Chapter 4), only five, all involving the oxides of nitrogen, are known:

EXAMPLE:

$$2NO + H_2 \rightarrow H_2O + N_2O.$$

In solution they are a little more numerous.

EXAMPLE:

$$2Fe^{3+} + Sn^{2+} \rightarrow 2Fe^{2+} + Sn^{4+}.$$

All third-order reactions known up until now correspond to the schema-type:

$$2A + B \rightarrow \text{Products}$$

The corresponding rate equation is $dx/dt = k(a-2x)^2 \times (b-x)$ having the general solution

$$kt = \frac{1}{(2b-a)^2}\left[\frac{(2b-a)2x}{a(a-2x)} \text{ Log } \frac{b(a-2x)}{a(b-x)}\right]$$

Remark: In experimental practice, the expressions may be simplified by the judicious choice of initial quantities. Thus for the third-order reaction above, if the initial concentration of A is double that of B, then

$$\frac{dx}{dt} = k'(a-x)^3$$

leading to the solution

$$k't = \frac{x(2a-x)}{2a^2(a-x)^2}.$$

For a reaction of nth order, $nA \rightarrow$ Products or $A_1 + A_2 + \cdots + A_n \rightarrow$ Products, if we use equal initial quantities of reactants, we obtain

$$kt = \frac{1}{n-1}\left[\frac{1}{(a-x)^{n-1}} - \frac{1}{a^{n-1}}\right]$$

in which the rate coefficient has the dimensions of $time^{-1}$ $concentration^{1-n}$.

1.3. Experimental Determination of the Order
of a Reaction

This determination yields the empirical rate law, which gives information on the mechanism of the reaction.

For simple reactions, the estimation of the order and the simultaneous calculation of the rate coefficient present no difficulties; on the other hand for complex reactions, this determination may become very difficult.

The following different methods are used to determine the order of a reaction.

A. *Method of Integration*

This procedure by trial and error consists in calculating the numerical values of the rate constant k, by substituting experimental results a, b, c, etc. in the algebraic expressions given above, and noting the order for which the values of k thus calculated are reasonably constant.

The following graphic variation is also used: the results are plotted as different functions of time, which will be linear when the order chosen corresponds to that of the reaction.

EXAMPLE:

$$\log C = f(t) \rightarrow \text{first order}$$
$$\frac{1}{C} = f(t) \rightarrow \text{second order}$$
$$\frac{1}{C^2} = f(t) \rightarrow \text{third order}.$$

This procedure is only applicable to reactions involving only a single reactant (or to particular cases), but enables us to confirm that the order remains constant with time.

Experimental results are never exact, and the experimenter faces a subjective choice in drawing a straight line through scattered points. In the same way, in the analytical procedure it is necessary to use an ordinary statistical procedure to obtain an 'average' for the constant k which indicates the possible error.

B. *Use of Half-Reaction Times*

For reactions which involve only a single reactant (or when equal initial quantities of each reactant are used), the time of half-reaction is given by

$$t_{1/2} = \frac{\text{constant}}{a^{n-1}}$$

where a is the initial concentration and

$$\text{constant} = \frac{2^{n-1} - 1}{(n-1)\,k}.$$

A series of experiments utilizing initial concentrations $a_1, a_2, a_3, \ldots$ is carried out, and the order is calculated by use of the formula

$$n = 1 + \log \frac{t_1/t_2}{a_2/a_1}.$$

The method is applicable to any other fractional duration of the reaction ($t_{1/3}$ for example), and allows us to determine fractional orders.

C. *Tangent Method*

Measurements of the rate of a reaction at successive time intervals may also be used. The order can then be obtained (if the reaction admits of one) by the expression

$$n = \frac{\log \mathscr{V}_1 - \log \mathscr{V}_2}{\log C_1 - \log C_2}$$

since the rate is related to the concentration by $\mathscr{V} = kC^n$.

This method is applicable to all reactions, simple and complex, and is of particular interest at the moment when $t = 0$ (initial rate), when the initial order is obtained; it is often easier to find the initial rate law in the first 10% of the reaction, before products which can influence the rate (reversible reactions, autocatalysis, inhibition, polymerizations, etc.) have been formed in sufficient concentration.

However, the measurement of an instantaneous rate is very difficult because it corresponds to a derivative and requires measurement of the slope of a tangent; there is no rigorous procedure for obtaining the rate from experimental data, and the result is always approximate (graphic measurement of the slope of the tangent with the aid of a mirror; calculation of ratios x/t, etc.). The best procedure is to calculate the empirical equation of the curve $x = f(t)$ during the first 10–20% extent of the reaction, using the expression $x = mt + nt^2 + pt^3$. The initial rate $\mathscr{V}_0 = m$ is obtained from three measurements of x carried out after three equal time intervals.

By repeating the experiments with different initial concentrations of the chosen reactant we obtain the initial order of the reaction.

D. *Isolation Method (Ostwald)*

When the reaction involves several reactants, it is convenient to determine the partial orders with respect to each reactant by a judicious choice of reactions conditions.

The reactant chosen (the limiting reactant), is in very small amount compared with the other reactants, whose concentrations accordingly are effectively constant throughout the reaction. The overall order of the reaction then becomes the order with respect to the limiting reactant (degeneracy), and is obtained by one of the preceding methods.

In effect, for a reaction having the rate

$$\mathscr{V} = k (a - x)^p (b - x)^q \dots$$

if the initial concentration $b, \dots \gg a$ (at least twenty fold) x can only vary from 0 to a, and is always negligible with regard to $b \dots$, so that

$$\mathscr{V} \# k' (a - x)^p \quad \text{with} \quad k' = kb^q.$$

The experiment can then be repeated with $b \ll a \dots$, in order to obtain the partial order q.

Remarks: (1) The isolation method can be applied to initial orders because at $t = 0$, $x = 0$ and $\mathscr{V}_0 = ka^p b^q$. By leaving b constant and varying a, p may be determined rapidly.

(2) In solution, the molecules of solvent are in large excess over the molecules of solute, so that it is often difficult to determine by kinetics whether they participate in the reaction.

1.4. Complex Reactions

These comprise the three following groups: reversible reactions, competitive reactions, and successive reactions. We shall describe them and indicate the modern methods of calculation used by the kineticist. The results thus obtained will be of great value in studying actual reactions and their detailed mechanisms.

A. *Reversible Reactions*

The forward reaction frequently shows initially a simple order, but this does not extend beyond about 10% reaction, because the products react in the reverse direction and the reaction appears to slow down. From the point of view of the reaction masses, the overall reaction is limited and comes more or less slowly to a state of 'chemical equilibrium'; kinetically, it then does not change with time.

The kinetics of the establishment of chemical equilibrium are very important in current practice, where the three following cases are most often met.

(a) *Forward and reverse reactions of first order.*

Type: $A \underset{k_2}{\overset{k_1}{\rightleftarrows}} B$.

The rate equation

$$\frac{dx}{dt} = k_1'(a - x) - k_2(b + x)$$

(a: initial concentration of A; b: initial concentration of B) can be integrated directly with the solution

$$(k_1 + k_2) t = \text{Log} \frac{k_1 a - k_2 b}{k_1 a - k_2 b - x (k_1 + k_2)}.$$

However, a better method of calculation, applicable to the following cases, consists in considering the conditions at equilibrium rather than initially.

EXAMPLE:

$$A \rightleftarrows B \quad t = 0 \quad |A| = a \quad |B| = 0$$

$$\frac{dx}{dt} = k_1 (a - x) - k_2 x.$$

If

$$t \to \infty, \quad \mathscr{V} \to 0, \quad x \to x_\infty$$

so that

$$k_1 (a - x_\infty) - k_2 x_\infty \to 0$$

$$\to k_1 + k_2 = k_1 \frac{a}{x_\infty}.$$

hence

$$\frac{dx}{dt} = k_1 (a - x) - k_1 \frac{(a - x_\infty)}{x_\infty} x$$

$$\frac{dx}{dt} = \frac{k_1 a \left(x_\infty - x \right)}{x_\infty}$$

which on integration gives

$$\left(k_1 + k_2 \right) t = \text{Log} \frac{x_\infty}{x_\infty - x}.$$

Such a reaction is kinetically equivalent to a simple first-order reaction, with the overall constant equal to the sum of the constants for the forward and the reverse reaction, and the concentration at equilibrium replacing the initial concentration.

Remark: (1) At equilibrium the constant K of the mass action law is given by:

$$\frac{k_1}{k_2} = \frac{x_\infty}{a - x_\infty} = \frac{|B_{eq}|}{|A_{eq}|} = K.$$

If K is determined by an analytical or thermodynamic procedure, only a simple kinetic experiment is necessary to calculate k_1 and k_2.

(2) The preceding laws have been tested experimentally and found to apply exactly to the following reactions: the mutarotation of glucose: glucose $\alpha \leftrightarrows$ glucose β, which can be followed in a polarimeter, and the enolization of ketones:

$$RCH_2COCH_3 \rightleftarrows RCH = C-CH_3$$
$$|$$
$$OH$$

(b) *Forward and reverse reactions of second-order.*

Type $A + B \leftrightarrows C + D$.

If at

$$t = 0, \quad |A| = a, \quad |B| = b, \quad |C| = c, \quad |D| = d,$$

the rate equation may be written:

$$\mathscr{V} = k_1 (a - x)(b - x) - k_2 (c + x)(d + x)$$

the integration of which is difficult. By using conditions at equilibrium, the calculations are simplified.

EXAMPLE:

$$t \rightarrow 0, \quad c = d = 0, \quad a = b$$
$$t \rightarrow \infty, \quad x \rightarrow x_\infty, \quad \mathscr{V} = 0 = k_1 (a - x_\infty)^2 - k_2 x_\infty^2.$$

On substituting for k_2 in the rate equation, we obtain

$$\frac{dx}{dt} = k_1 (a - x)^2 - \frac{(a - x_\infty)^2}{x_\infty} x^2$$

and on rearranging

$$\frac{dx}{dt} = k_1 \frac{a}{x_\infty^2} (2x_\infty - a) \left(\frac{a x_\infty}{2x_\infty - a} - x \right)(x_\infty - x)$$

which is analogous to the second-order equation

$$\mathscr{V} = k' (a' - x)(b' - x)$$

and, like it, is solved as

$$k_1 t = \frac{x_\infty}{2a(a - x_\infty)} \operatorname{Log} \frac{x(a - 2x_\infty) + a x_\infty}{a(x_\infty - x)}.$$

Remark: For reactions of the type $2A \leftrightarrows C+D$ (ex.: $2IH \leftrightarrows I_2 + H_2$), the rate equation is given by

$$t \to 0, \quad |IH| = C$$

$$-\frac{1}{2}\frac{d\,|IH|}{dt} = \frac{1}{2}\frac{dy}{dt} = k_1 (c-y)^2 - k_2 \left(\frac{y}{2}\right)^2 .$$

If one knows the equilibrium constant for formation of HI ($K = k_2/k_1$) one can obtain k_1 by the equation

$$k_1 = \frac{1}{2cKt}$$

$$\left\{ \mathrm{Log}\; \frac{\left[c\left(1+\sqrt{\frac{K}{4}}\right) \middle/ 1 - \frac{K}{4}\right] - y}{\left[c\left(1-\sqrt{\frac{K}{4}}\right) \middle/ 1 - \frac{K}{4}\right] - y} - \mathrm{Log}\; \frac{1+\sqrt{\frac{K}{4}}}{1-\sqrt{\frac{K}{4}}} \right\} .$$

The same method of calculation may be applied to the reverse (synthetic) reaction. On going from a moles of H_2 and b moles of I_2, we obtain

$$\frac{1}{2}\frac{d\,(IH)}{dt} = \frac{1}{2}\frac{dx}{dt} = k_2 \left(a - \frac{x}{2}\right)\left(b - \frac{x}{2}\right) - k_1 x^2$$

from which we can obtain k_2 by kinetics, independently of the value of k_1. When this is done, the values of K (analytic), and of k_1 and k_2 obtained from two different series of experiments prove to be compatible within reasonable limits of precision.

(c) *Forward reaction of second-order, and reverse reaction first order.*

Type $A + B \underset{R_2}{\overset{R_1}{\rightleftharpoons}} C$.

If at $t=0$, $|A| = |B| = a$, $|C| = 0$, the rate equation becomes

$$\frac{\mathrm{d}x}{\mathrm{d}t} = k_1 (a - x)^2 - k_2 x$$

having the following solution, obtained by the method already indicated

$$k_1 t = \frac{x_\infty}{a^2 - x_\infty^2} \mathrm{Log} \frac{x_\infty (a - x x_\infty)}{a^2 (x_\infty - x)}.$$

EXAMPLE: Isomerisation of an ammonium cyanate to give a substituted urea:

$$CNO^- + RNH_3^+ \rightleftharpoons CO\!\!\!\overset{\displaystyle NHR}{\underset{\displaystyle NH_2}{<}}$$

B. *Successive Reactions*

Successive reactions take place in steps, with the formation of intermediate products. The steps may be irreversible: the starting material and the different intermediates react in one direction to give the product of the reaction. Alternatively, one or more steps may be reversible, or again parallel, with formation of several products (competitive reactions).

FIRST EXAMPLE. General case, n successive reactions of first or second order.

A reactant A having an initial concentration C_0 which disintegrates spontaneously (a radioactive sequence) or which reacts with a substance B, present at a concentration $b \gg C_0$ (degeneracy) to give a stable product A_n. In the latter case, the intermediate products may also react with B.

$$\begin{array}{cccc} & \overset{k_1}{\rightarrow} A_1 & \overset{k_2}{\rightarrow} A_2 & \overset{k_n}{\rightarrow} \ldots A_n \end{array}$$

$$\left. \begin{array}{cccc} t = 0 & C_0 & O & O & O \\ t & C & C_1 & C_2 & C_n \end{array} \right\} \text{first-order disintegration}$$

In the second case:

$$\begin{array}{ccccc} A & \overset{k'_1}{\rightarrow} A_1 & \overset{k'_2}{\rightarrow} A_2 & \overset{k'_n}{\rightarrow} \cdots \rightarrow A_n. \\ + & + & + \\ B & B & B \end{array}$$

Each step follows a second-order law, but the condition that $b \gg C_0$ relates this case to the previous one since

$$k_1 = k'_1 b$$
$$k_2 = k'_2 b$$
$$k_n = k'_n b.$$

The rate equations for each step are as follows

(1) $\dfrac{dC}{dt} = -k_1 C;$

(2) $\dfrac{dC_1}{dt} = k_1 C - k_2 C_1;$

(3) $\dfrac{dC_2}{dt} = k_2C_1 - k_3C_2;$

(n) $\dfrac{dC_n}{dt} = k_nC_{n-1}.$

The solution of such a system of differential equations can be effected by the aid of operational calculus. By replacing each differentiation d/dt by a multiplication using the corresponding operator P, the preceding equations become

$$\frac{dC}{dt} = \frac{d(C-C_0)}{dt} = PC - PC_0$$
$$PC_1 = k_1C - k_2C_1$$
$$PC_2 = k_2C_1 - k_3C_2$$
$$PC_{n-1} = k_{n-1}C_{n-2} - k_nC_{n-1}$$
$$PC_n = k_nC_{n-1}.$$

By treating the operator P like an ordinary algebraic variable, we obtain

$$C = \frac{PC_0}{P+k_1} \qquad C_1 = \frac{k_1PC_0}{(P+k_1)(P+k_2)}$$
$$C_2 = \frac{k_1k_2PC_0}{(P+k_1)(P+k_2)(P+k_3)}$$
$$C_{n-1} = \frac{k_1k_2\ldots k_{n-1}PC_0}{(P+k_1)\ldots(P+k_n)}$$
$$C_n = \frac{k_1k_2\ldots k_nC_0}{(P+k_1)\ldots(P+k_n)}.$$

By using presently available tables of Laplace transformations, we can replace the transformations or the unknowns $C, C_1, \ldots C_n$ by their values and obtain

$$C = C_0 e^{-k_1 t}$$

$$C_1 = C_0 \left[\frac{k_1}{k_2 - k_1} e^{-k_1 t} + \frac{k_1}{k_1 - k_2} e^{-k_2 t} \right]$$

$$C_2 = C_0 \left[\frac{k_1 k_2}{(k_2 - k_1)(k_3 - k_1)} e^{k_1 t} + \frac{k_1 k_2}{(k_1 - k_2)(k_3 - k_2)} \right.$$
$$\left. \times e^{-k_2 t} + \frac{k_1 k_2}{(k_1 - k_3)(k_2 - k_3)} e^{-k_3 t} \right]$$

$$C_{n-1} = C_0 \left[\frac{k_1 k_2 \ldots k_{n-1}}{(k_2 - k_1)(k_3 - k_1) \ldots (k_n - k_1)} e^{-k_1 t} \right.$$
$$+ \frac{k_1 k_2 \ldots k_{n-1}}{(k_1 - k_2)(k_3 - k_2) \ldots (k_n - k_2)} e^{-k_2 t} + \cdots$$
$$+ \frac{k_1 k_2 \ldots k_{n-1}}{(k_1 - k_3)(k_2 - k_3) \ldots (k_n - k_3)} e^{-k_3 t}$$
$$\left. + \frac{k_1 k_2 \ldots k_{n-1}}{(k_1 - k_n)(k_2 - k_n) \ldots (k_{n-1} - k_n)} e^{-k_n t} \right]$$

$$C_n = C_0 \left[I - \frac{k_2 k_3 k_4 \ldots k_n}{(k_2 - k_1)(k_3 - k_1) \ldots (k_n - k_1)} e^{-k_1 t} \right.$$
$$- \frac{k_1 k_3 k_4 \ldots k_n}{(k_1 - k_2)(k_3 - k_2) \ldots (k_n - k_2)} e^{-k_2 t} - \cdots$$
$$\left. - \frac{k_1 k_2 \ldots k_{n-1}}{(k_1 - k_n)(k_2 - k_n) \ldots (k_{n-1} - k_n)} e^{-k_n t} \right]$$

$$= C_0 - [C + C_1 + C_2 + \cdots + C_{n-1}].$$

SECOND EXAMPLE. Case of two successive steps.

This case can be treated directly without the use of the preceding formulae.

$$\begin{array}{cccc} & A & \xrightarrow{k_1} & B & \xrightarrow{k_2} & C \\ t = 0 & a & & 0 & & 0 \\ t & a - x & & x - y & & y \end{array} \Bigg\} \text{ conditions.}$$

For the first step

$$\frac{dx}{dt} = k_1 (a - x) \rightarrow x = a (1 - e^{-k_1 t}).$$

For the second step

$$\frac{dy}{dt} + k_2 y = k_2 x = k_2 a (1 - e^{-k_1 t}).$$

The solution of this two-membered differential equation consists in adding a particular solution of the two-membered equation to a general solution of the equation without the second member.

The particular solution is

$$y = M + N e^{-k_1 t}$$

with M and N constants determined by identification.

The general solution is

$$y = P e^{-k_2 t}.$$

Since

$$\frac{dy}{dt} = - k_1 N e^{-k_1 t}$$

this replaced in the differential equation gives

$$-k_1 N e^{-k_1 t} + k_2 (M + N e^{-k_1 t}) = k_2 (1 - e^{-k_1 t}) a$$

or

$$e^{-k_1 t} (-k_1 N + k_2 N + a k_2) = a k_2 - k_2 M$$

which holds for all values of t (including $t = 0$). This requires

$$M = a \quad \text{and} \quad N = \frac{a k_2}{k_1 - k_2}.$$

Similarly, from a consideration of initial conditions $t = 0$ and $y = 0$,

$$P = \frac{a k_1}{k_2 - k_1}.$$

Finally, since $|B| = x - y$ it follows that

$$|A| = a e^{-k_1 t}$$

$$|B| = \frac{a k_1}{k_2 - k_1} \left[e^{-k_1 t} - e^{-k_2 t} \right]$$

$$|C| = a \left[1 - \frac{k_2}{k_2 - k_1} e^{-k_1 t} + \frac{k_1}{k_2 - k_1} e^{-k_2 t} \right]$$

The concentration of B passes through a maximum $d|B|/dt = 0$, which requires

$$|B|_{max} = a \left(\frac{k_1}{k_2} \right)^{k_2/(k_2 - k_1)}$$

at the time

$$t_{max} = \frac{\text{Log } k_1/k_2}{k_1 - k_2}.$$

The maximum is the more accentuated, the more the second step is slower than the first ($k_2 < k_1$), and corresponds to a point of inflexion in curve $C (d^2C/dt^2 = 0)$.

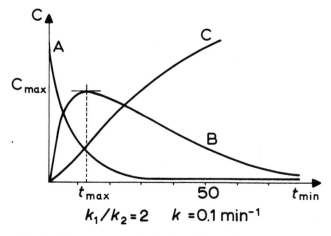

Fig. 1.   Curves representing A, B, and C as a function of time.

If $k_1 \ll k_2$, the intermediate B reacts very rapidly and cannot accumulate. It then becomes difficult to demonstrate the presence of B experimentally: this explains the long period of confusion over whether certain reactions proceeded by one or two steps.

THIRD EXAMPLE. Two steps, consecutive and reversible.

$$A \underset{k_1'}{\overset{k_1}{\rightleftarrows}} B \underset{k_2'}{\overset{k_2}{\rightleftarrows}} C$$

$$\begin{cases} t = 0 \\ t \end{cases} \begin{cases} A_0 & O & O \\ A & B & C \end{cases} \text{concentrations}$$

$$\frac{dA}{dt} = -k_1 A + k_1' B$$

$$\frac{dB}{dt} = k_1 A - k_1' B - k_2 B + k_2' C$$

$$\frac{dC}{dt} = k_2 B - k_2' C.$$

These functions are then transformed

$$PA - PA_0 = k_1' B - k_1 A$$
$$PB = k_1 A - k_1' B - k_2 B + k_2' C$$
$$PC = k_2 B - k_2' C$$

which leads to

$$A = \frac{A_0 \left[P^2 + P(k_1' + k_2 + k_2') + k_1' k_2' \right]}{(P + \gamma_1)(P + \gamma_2)}$$

$$B = \frac{A_0 k_1 (P + k_2')}{(P + \gamma_1)(P + \gamma_2)}$$

$$C = \frac{A_0 k_1 k_2}{(P + \gamma_1)(P + \gamma_2)}$$

where γ_1 and γ_2 are the roots of opposite sign of the equation

$$\gamma^2 + \gamma(k_1 + k_2 + k_1' + k_2') + k_1 k_2 + k_1 k_2' + k_1' k_2' = 0.$$

By using operational calculus tables, we obtain

$$A = A_0 \left[\frac{k_1' k_2'}{\gamma_1 \gamma_2} + \frac{\gamma_1^2 - \gamma_1 (k_1' + k_2' + k_2) + k_1' k_2'}{\gamma_1 (\gamma_1 - \gamma_2)} e^{-\gamma_1 t} \right.$$
$$\left. + \frac{\gamma_2^2 - \gamma_2 (k_1' + k_2' + k_2) + k_1' k_1'}{\gamma_2 (\gamma_2 - \gamma_1)} e^{-\gamma_2 t} \right]$$

$$B = k_1 A_0 \left[\frac{k_2'}{\gamma_1 \gamma_2} + \frac{k_2' - \gamma_1}{\gamma_1 (\gamma_1 - \gamma_2)} e^{-\gamma_1 t} + \frac{k_2' - \gamma_2}{\gamma_2 (\gamma_2 - \gamma_1)} e^{-\gamma_2 t} \right]$$

$$C = k_1 k_2 A_0 \left[\frac{1}{\gamma_1 \gamma_2} + \frac{1}{\gamma_1 (\gamma_1 - \gamma_2)} e^{-\gamma_1 t} + \frac{1}{\gamma_2 (\gamma_2 - \gamma_1)} e^{-\gamma_2 t} \right]$$

As $r \to \infty$:

$$A_\infty \to A_0 \frac{k_1' k_2'}{\gamma_1 \gamma_2}$$

$$B_\infty \to A_0 \frac{k_1 k_2'}{\gamma_1 \gamma_2}$$

$$C_\infty \to A_0 \frac{k_1 k_2}{\gamma_1 \gamma_2}$$

and the system is then in a state of equilibrium (triangular equilibrium).

C. *Parallel Reactions*

These reactions are divided into two distinct subgroups:
Twinned reactions of type

$$A + B \begin{cases} \nearrow C \\ \searrow D \end{cases}$$

and concurrent or competing reactions

$$
A \underset{+C \searrow F}{\overset{+B \nearrow D}{}}
$$

D. *Twinned Reactions*

These reactions are often encountered in organic chemistry, where the yield very rarely reaches 100% because of the formation of by-products. For example, the chorination of toluene gives chiefly the *ortho* and *para* isomers, with a very small amount of the *meta* isomer.

EXAMPLE: The attack of hydroxide ion on an alkyl halide can result in both substitution and elimination reactions

$$
RCH_2CH_2Br + OH \begin{cases} RCH_2CH_2OH & \text{(substitution)} \\ RCH = CH_2 & \text{(elimination)} \end{cases}
$$

The overall order of the two reactions is often identical; which facilitates the kinetic analysis. In such a case, if x is the quantity of reactant A which has been consumed, y and z the quantities of the products C and D which have been formed at time t, and a and b are the initial concentrations of A and B, then

$$
\mathscr{V}_C = \frac{dy}{dt} = k_1 (a - x)^m (b - x)^n
$$

$$\mathscr{V}_D = \frac{\mathrm{d}z}{\mathrm{d}t} = k_2 (a - x)^m (b - x)^n$$

$$\mathscr{V}_A = \frac{\mathrm{d}x}{\mathrm{d}t} = \frac{\mathrm{d}y}{\mathrm{d}t} + \frac{\mathrm{d}z}{\mathrm{d}t} = (k_1 + k_2) (a - x)^m (b - x)^n.$$

Such a system becomes equivalent to a simple system of $(m + n)$th order. For example, if $m = n = 1$,

$$\frac{\mathrm{d}x}{\mathrm{d}t} = (k_1 + k_2) (a - x) (b - x)$$

$$\rightarrow (k_1 + k_2) t = \frac{1}{a - b} \mathrm{Log} \frac{b (a - x)}{a (b - x)}$$

Furthermore, at every instant t

$$\frac{\mathrm{d}y}{\mathrm{d}z} = \frac{k_1}{k_2}, \quad y = \frac{k_1}{k_2} z.$$

The relative proportion of the two products C and D is independent of time. For $t \rightarrow \infty$, the proportion of the two products, determined by chemical analysis, gives the ratio of the two rate constants.

Remarks: (1) In all kinetic studies, care should be taken to establish that the reaction studied is the only one taking place; it is not sufficient to study the disappearence of a reactant if it is not certain, by chemical analysis, that by-products are not being formed. In general, whenever possible it should be shown that the rate of formation of a product is equal to the rate of disappearance of a reactant. Too often, the first is slightly lower, indicating the presence of a parallel reaction.

(2) In the case when the partial order with respect to one of the two reactants A and B is not the same in the two twinned reactions, it is preferable (following a previous example) to choose experimental conditions such that $b \gg a$. The two apparent constants will then be $k'_1 = k_1 b$ and $k'_2 = k_2 b$, calculated from the equation

$$(k'_1 + k'_2)\, t = \text{Log}\, \frac{a}{a - x}.$$

E. *Competing Reactions*

Type $A \Big\langle \substack{+B \nearrow D \\ +C \searrow F}$

This case is very important practically in industry, but has been the object of only a few detailed studies, because refined experimental methods are often required, and the mathematical treatment of the problem often proves very difficult, even using the most sophisticated procedures.

Also, kineticists often adjust their experimental conditions to diminish the complexity of the reaction. In particular, if the concentrations of the two reactants B and C are made much higher than that of the common reactant A, the competing reactions degenerate to twinned reactions, which are much easier to deal with, especially if the partial order with respect to A is 1 or 2. Similarly, if the concentration of A is very much greater than that of B and C, the two reactions become independent

$$B \overset{+A}{\to} D \quad k'_D = k_D |A|$$

$$C \overset{+A}{\to} F \quad k'_F = k_F |A| .$$

In general, it is necessary for the order of the two reactions with respect to A to be the same if the mathematical treatment is to be much simplified.

If a, b, and c are the initial concentrations of A, B, and C at $t=0$, and x, y, and z the quantities which have been consumed at time t, then

$$-\frac{d|B|}{dt} = \frac{d|D|}{dt} = k_D (a - x)^q (b - y)^m$$

$$-\frac{d|C|}{dt} = \frac{d|F|}{dt} = k_F (a - x)^q (c - z)^n$$

and

$$\frac{dx}{dt} = k_D (a - x)^q (b - y)^m + k_D (a - x)^q (c-z)^n.$$

Even in the case when $q=m=n=1$, so that the two competing reactions are both second order, the concentrations of A, B and C as a function of time are expressed in the form of an infinite series of converging terms, which is difficult to solve in practice.

On the other hand, by eliminating the time factor, it is possible to obtain analytically the necessary relations. Thus, if $m=n=1$, the relation:

$$\frac{dy}{dz} = \frac{k_D}{k_F} \frac{b - y}{c - z}$$

gives by integration (taking account of the constant of integration)

$$\text{Log} \frac{b}{b-y} = \frac{k_D}{k_F} \text{Log} \frac{c}{c-z}$$

from which a straight-line curve of slope k_D/k_F may be obtained.

If $m=n=2$, we obtain

$$\frac{dy}{dz} = \frac{k_D (b-y)^2}{k_F (c-z)^2} \rightarrow \frac{1}{b-y} - \frac{1}{b} = \frac{k_D}{k_F}\left(\frac{1}{c-z} - \frac{1}{c}\right)$$

and by analogy, if $m=1$, $n=2$

$$\text{Log} \frac{b}{b-y} = \frac{k_D}{k_F}\left(\frac{1}{c-z} - \frac{1}{c}\right).$$

If $m=1$, $n=0$ (as frequently happens in solution when the solvent intervenes in the reaction):

$$\frac{dy}{dz} = \frac{k_D}{k_F}(b-y) \quad dx = dy + dz = dy + \frac{k_F}{k_D}\frac{dy}{b-y}$$

$$x = y + \frac{k_F}{k_D} \text{Log} \frac{b}{b-y}.$$

These different examples allow us to compare the relative reactivities of two substances for a common reactant, but the procedure is used only if one of the two systems cannot be examined independently (as in the case of very rapid reactions). Thus, if the system A $\xrightarrow{+B}$ D is too rapid for the experimental method being used, we can add to B another reactant C, before adding the mixture to A. The competitive reactions which then ensue enable us to deduce k_D, if k_F (for the reaction or C) is known from some independent experiment.

Kinetic studies of competitive reactions have resulted in the development of separation and purification procedures by 'fractional reaction', by making use of the relative reactivity of a reactant with respect to two substrates. An example is the preparation of spectroscopically pure (optically transparent) cyclohexane, free from benzene, by selective chlorination of the latter.

CHAPTER 2

KINETIC THEORIES OF ELEMENTARY REACTIONS

A. *Elementary Reactions and Actual Reactions*

An elementary reaction corresponds to a monomolecular, bimolecular or trimolecular process which does not involve any intermediate.

Actual reactions are in general complex, and the elucidation of their mechanism consists in finding the number of elementary steps, successive or parallel, by which reactants go over into products.

The mechanism of an elementary step implies one of the following simple processes, following the localization of vibrational energy and its transformation into translational energy:

(1) Rupture or creation of a simple chemical bond.

(2) Rupture of a bond accompanied by the formation of another bond.

(3) Simultaneous rupture and formation of two bonds (a rather rare process, corresponding to the so-called four-centre reactions).

These considerations gave rise to the principle of least structure change by Müller (1856):

"All chemical change takes place by a series of elementary reactions, each of which involves only a minimum change."

On the basis of the kinetics of equilibrium reactions, we can add to this principle that of 'specific microreversibility'.

"In a reversible reaction, the path followed by the reaction in one direction is in all details that followed by the reaction in the reverse direction."

In practice it is often difficult to be certain that a given reaction is elementary, even if it follows one of the simple kinetic laws of first, second or third order, and this difficulty is reflected in the confusion between the notions of order and of molecularity. It is enough for one of the elementary steps to be clearly slower than the others for it to become the rate-determining stage and to determine the overall rate.

Writing a reaction as elementary represents a molecular fact, and does not necessarily indicate an observed rate.

Numerous theories have been proposed to take account of the mechanism of elementary reactions. At present transition state theory is the most satisfactory, and enables us to explain the failings of older theories such as elementary collision theory and the theory of monomolecular reactions, which have still practical usefulness.

B. *Transition State Theory*

In practice, this term has become synonymous with 'absolute rate theory' or 'the theory of the activated complex'.

The rearrangement of atoms in a chemical reaction takes place by the movement of atomic nuclei in the electronic cloud; this corresponds in space to movement on a potential energy surface, on which the path leading from reactants to products (and *vice-versa*) will be that of lowest energy.

The calculation of such a surface represents a particularly difficult problem in quantum mechanics, and in spite of the advent of computers has been solved only for a small number of simple cases such as

$$H^{\cdot} + H_2 \rightarrow H_2 + H^{\cdot}.$$

If we represent the variation of potential energy along the reaction path as a function of the extent of the reaction

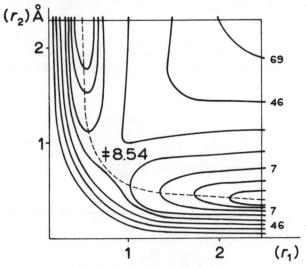

Fig. 2. Potential energy surface.

(the 'reaction coordinate') on a two-dimensional diagram, we obtain Figure 3.

The energy maximum corresponds to the transition state (the active complex) where the rearrangement of the atomic nuclei and the electrons is partially completed, and which has a finite lifetime.

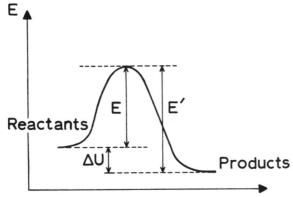

Fig. 3. Diagram of reaction path, showing energy of activation E.

PRECEDING EXAMPLE:

$$[H - \underset{r_1}{-} - H - \underset{r_2}{-} - H]$$

in which the distances r_1 and r_2 are greater than in the molecule H_2.

EXAMPLE: nucleophilic substitution

$$OH^- + CH_3Br \rightarrow CH_3OH + Br^-$$

where the atoms of carbon and hydrogen approach co-planarity in the active complex.

This active complex, corresponding to a maximum on a potential energy surface, should not be confused with an intermediate (a real chemical entity corresponding to an energy minimum) because one of its vibrations is replaced by a translation along the reaction coordinate.

The difference in energy between the reactants and the active complex constitutes a potential barrier or energy of activation E. If the reaction is reversible, the reverse reaction has a barrier E', which will be greater if the reaction is exothermic.

These two energies of activation are related to the heat of reaction (at constant volume) by the formula

$$\Delta U = E - E'.$$

The rate of the chemical reaction $A + B \rightarrow$ Products is equal to the number of activated complexes passing over the activation barrier each second, which is given by the concentration of activated complexes multiplied by the transmission coefficient v, a universal factor for all systems given by the relation

$$v = \left. \frac{\mathscr{K}T}{h} \right\} \quad \begin{array}{l} h = \text{Planck's constant.} \\ \mathscr{K} = \text{Boltzman's constant.} \end{array}$$

and of the order of 10^{13} s^{-1} at 25°C.

Even although the activated complex corresponds to an energy maximum, we can consider it to be in equilibrium with the reactants and apply the mass action law

$$A + B \rightleftarrows (A - - B)^{\neq} \rightarrow \text{products}$$

with

$$|A - - B|^{\neq} = K^{\neq} |A| |B|$$

and

$$K^{\neq} = e^{-\varDelta G_0 \neq /RT} = |e^{-\varDelta H_0 \neq /R\iota} e^{\varDelta S_0 \neq /R}.$$

$\varDelta G_0^{\neq}$, $\varDelta H_0^{\neq}$, $\varDelta S_0^{\neq}$ are the standard changes in free energy of activation, enthalpy of activation, and entropy of activation for the reaction $A + B \rightarrow (A - - - B)^{\neq}$, so that the rate coefficient k can be expressed by the general formula tenable under all conditions

$$k = \frac{\mathscr{K}T}{h} e^{-\varDelta H_0 \neq /RT} e^{\varDelta S_0 \neq /R}.$$

Remarks: (1) Experimental equation of Arrhénius.

Arrhénius advanced the following empirical formula for rate constants:

$$k = Ae^{-E/RT} \quad \begin{array}{l} A \rightarrow \text{frequency factor} \\ E \rightarrow \text{activation energy.} \end{array}$$

At constant pressure, the variation of k with temperature according to the two formulae above is given by

$$\left(\frac{\delta \log k}{\delta T}\right)_P = \frac{\varDelta H_0^{\neq} + RT}{RT^2} \quad \left(\frac{\delta \log k}{\delta T}\right)_P = \frac{E}{RT^2}$$

so that $E_{\text{Arrhénius}} = \varDelta H_0^{\neq} + RT$. In practice, the term RT is often negligible, being less than 2 kcal mole^{-1}, so that even

over a large temperature range the empirical relationship
of Arrhénius holds well.

(2) The preceding relationship enables us to define more
exactly the nature of the energy barrier, which in the
general case corresponds to a free energy of activation
$\Delta G_0^{\neq}$. (Only in the case when $\Delta S_0^{\neq} = 0$, is the term, energy
of activation, strictly correct.)

The greater this energy barrier, the slower the reaction.
(At 15 °C, a reaction is rapid if ΔG_0 is less than 10 kcal
mole^{-1}, negligible if it is over 50 kcal mole^{-1}.) However,
even when the enthalpy of activation is high, the reaction
can proceed if the entropy of activation is sufficiently
large and positive: this corresponds to a compensation of
the energy factor by a favourable geometrical configuration.

(3) At the present time, it is not possible in general to
obtain the free energy of activation from numerical data
tabulated in the literature. The enthalpy of activation is
obtained experimentally by measuring the reaction rate
as a function of temperature (by a plot of Log k/T as a
function of $(1/T)$), and Arrhénius energy of activation by
a plot of log k as a function of $(1/T)$. The two plots should
be linear with slopes equal to

$$-\frac{\Delta H_0^{\neq}}{R} \quad \text{or} \quad -\frac{E}{R}$$

respectively.

In the same way, entropies of activation are rarely
available, although sometimes they can be deduced from

spectroscopic data. However, the following considerations enable us sometimes to obtain their order of magnitude.

For monomolecular reactions, the activated complex often has a structure almost identical with that of the resting molecule, so that $\Delta S_0^{\neq} = 0$. The frequency factor A is then almost identical to $\mathscr{K}T/h$, a universal frequency and so is roughly the same for all monomolecular processes.

For molecular addition processes, experience has shown the structure of the activated complex is often very close to that of the reaction product. The entropy of activation is then roughly equal to the entropy change in the reaction

$$A + B \rightleftarrows (A - - - B)^{\neq} \rightarrow AB$$
$$\Delta S_0^{\neq} = S_{AB}^{\neq} - S_A - S_B \,\#\, S_{AB} - S_A - S_B \,\#\, \Delta S_0 .$$

(4) The rate constants of chemical reactions increase strongly with temperature (thermal reactions); on the contrary, those of nuclear reactions are independent of temperature and do not change with time.

C. *Activated Complex and Reaction Intermediates*

When the reaction studied is not elementary, but consists of a succession of elementary steps, each of these (according to transition state theory) goes through a transition state which corresponds to a maximum in free energy of activation; each minimum between these maxima corresponds to an intermediate compound, in principle detectable by physico-chemical means, although the difficulties in doing this increase with the magnitude of the free energy of the minimum. However, the concentration of such intermediates can become significant in favourable cases. On the other hand, this is never the case for activated complexes because of the very small value of the constant $K^{\neq}$, of the order of 10^{13} K, so that $K^{\neq}$ never attains a perceptible value except in super-rapid reactions; however, the reaction is then over in a time of the order of molecular relaxations.

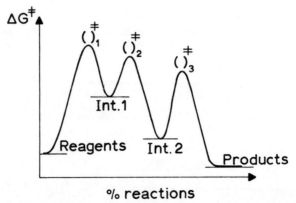

Fig. 4.   Free energies of activation and reactive intermediates.

Remark: If an intermediate cannot be recognized or followed by physico-chemical measurements, it is not possible to deduce from the experimental overall rate any information on the individual steps and on the constitution of the successive activated complexes.

D. *The Transition State and Elementary Collision Theory*

Elementary collision theory was formulated after the development of the kinetic theory of gases; although it has been superceded by transition state theory, it is frequently used because its limitations can be explained and even anticipated.

The number of collisions in a gaseous mixture of molecules of A with molecules of B is

$$z_{AB} = \pi\sigma_m^2 \left[\frac{8RT\,(M_A + M_B)}{\pi M_A\,M_B} \right]^{1/2} n_A n_B$$

with

$$\sigma_m = \frac{\sigma_A + \sigma_B}{2}$$

if one assumes the collision of two rigid spheres; σ_A and σ_B are then the collision diameters of each molecule, and the last factor represents the relative average velocity. In reality forces of attraction or repulsion between the molecules make the impact parameter b smaller or larger than the sum $((\sigma_A + \sigma_B)/2)$, or distance of closest approach, and the term πb_{max}^2 is termed the 'capture cross-section'.

The capture cross-sections of thermal reactions are of the order of molecular dimensions, that is to say

$$10^{-1}\ \text{Å}^2 < \pi\sigma_m^2 < 10^2\ \text{Å}^2,$$

except in the case of ionic reactions, where the electrostatic forces are effective at considerably greater distances than intermolecular forces, so that the capture cross-

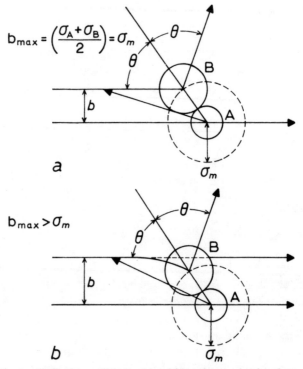

Fig. 5. Molecular collisions. (a) without intermolecular forces; (b) with attractive force.

section can be much greater. For nuclear reactions, the capture cross-sections correspond to the size of a nucleon and are expressed in barns (10^{-8} A^2 or 10^{-24} cm^2), and are thus 10^8 smaller than chemical cross-sections.

At ordinary temperatures collisions are very frequent, of

the order of 10^{28} s^{-1} cc^{-1}, so that all chemical reactions should be instantaneous if all of these collisions resulted in reaction. However, in fact only encounters between molecules having a kinetic energy greater than a characteristic value of E are effective; their number is given by the Boltzmann factor, $e^{-E/RT}$, which gives as the rate expression

$$\mathscr{V} = k n_A n_B = \pi \sigma_m^2 \left(\frac{8RT}{\pi} \frac{M_A + M_B}{M_A M_B} \right)^{1/2} n_A n_B \, e^{-E/RT}$$

so that

$$k = \pi \sigma_m^2 \left(\frac{8RT}{\pi} \frac{M_A + M_B}{M_A M_B} \right)^{1/2} e^{-E/RT}$$
$$= Z_{AB} e^{-E/RT}$$

Z_{AB} is the collision frequency, of the order of 10^{-10} cm^3 s^{-1} molecule^{-1} or

$$6 \times 10^{+10} \, l^{+1} \, \text{mol}^{-1} \, \text{s}^{-1} \left(x \frac{N}{10^3} \right).$$

Remarks: (1) For reactions taking place between two molecules of the same gas (2A → Products), $Z_{AA} = \pi \sigma^2 (4RT/\pi M)^{1/2}$ so that the rate coefficient is

$$k = -\frac{1}{2n_A^2} \frac{dn_A}{dt} = Z_{AA} \, e^{-E/RT}.$$

(2) Steric factor. The collision frequency Z_{AB} would be identical to the experimental frequency factor A of Arrhénius if the elementary theory were exact; however, while in a few cases these two quantities are of the same

order of magnitude, in general they differ by several powers of 10. This necessitates the introduction of a corrective term P (steric or probability factor), so that

$$A = PZ_{AB}.$$

On comparing this last formula with one obtained previously

$$A = \frac{\mathscr{K}T}{h} e^{\Delta S_0 \neq /R} = PZ_{AB}$$

the steric factor becomes equal to

$$\frac{\mathscr{K}T}{hZ_{AB}} e^{\Delta S_0 \neq /R},$$

and depends on the entropy of activation and hence on the geometrical configuration of molecules making up the activated complex. This theory thus gives results which can be used for reactions in the gaseous phase; however, its application to solutions is much less dependable.

E. *Monomolecular Reactions*

Transition state theory applies to all mono-, bi-, trimolecular processes, in the gas phase as well as in liquid phase (and equally to heterogeneous processes), in contrast to elementary collision theory, which is limited to bimolecular reactions in the gas phase.

In 1922 Lindeman produced a more satisfactory theory of monomolecular reactions, by insisting on the necessity of an activation process by collision between identical molecules. The mechanism was then the following:

$$A + A \underset{2}{\overset{1}{\rightleftarrows}} A^* + A$$
$$\downarrow^3 \; B + C$$

with

$$\frac{d|A^*|}{dt} = k_1 |A|^2 - k_2 |A||A^*| - k_3 |A^*|$$

$$-\frac{d|A|}{dt} = k_1 |A|^2 - k_2 |A^*| |A|$$

$$\frac{d|B|}{dt} = k_3 |A^*|.$$

This is only soluble by the steady-state approximation (see Chapter 4)

$$\frac{d|A^*|}{dt} \neq 0$$

leading to

$$|A^*| = \frac{k_1 |A|^2}{k_2 |A| + k_3} \quad \frac{d|B|}{dt} = \frac{k_1 k_3 |A|^2}{k_2 |A| + k_3}.$$

In the case when the rate of fragmentation 3 is very low compared to the deactivation 2 $(k_2 |A| \gg k_3)$,

$$\frac{d|B|}{dt} \rightarrow \frac{k_1 k_3}{k_2} |A| \rightarrow \quad \text{first-order reaction.}$$

When the opposite limiting case holds,

$$\frac{d|B|}{dt} \rightarrow k_1 |A|^2 \rightarrow \quad \text{second-order reaction.}$$

These last considerations allow one to verify the theory of Lindeman. In the gas phase, it is found that the rate constants of first-order reactions diminish when the pressure drops below 10 mm Hg: the probability of deactivation 2 is then much diminished. The injection of an inert gas then causes the value of the rate constant to rise again. The same phenomenon is not produced in solution by dilution, because the deactivation 2 is brought about by molecules of solvent S:

$$A^* + S \underset{2}{\rightarrow} A + S.$$

This theory of monomolecular reactions has been recently further developed by Slater (1959).

CHAPTER 3

EXPERIMENTAL METHODS

The methods of preparing or manufacturing chemical products are divided into two fundamentally opposite procedures: static and dynamic. In the same way, kinetic methods can be either transient or stationary.

3.1. Static Procedure

This discontinuous process is widely used, even in industrial reactors; it consists of mixing the reactants at a time $t = 0$, and homogenizing constantly the reaction mixture. It is possible to operate at a constant pressure or volume, isothermally, etc. The system is thermodynamically closed, the concentrations of the different reactants and products vary continually with time: hence a transient operation.

It is not possible to measure directly the rate of the reaction $\mathscr{V}$, which must be deduced from the variation of concentrations with time. These variations can be obtained by the aid of any of the non-destructive physicochemical methods of analysis (continuous measurement of pressure, volume, index of refraction, optical density, conductivity, specific inductive power, pH, electrode potential, etc.) or even, in the case of reactions sufficiently slow, by destructive chemical analysis of aliquots of a large volume of the reaction mixture (if necessary, after the reaction has been stopped by drastic chilling of the aliquot).

3.2. Dynamic Procedure

However, continuous processes are better adapted for the usual industrial conditions of operation, as well as for kinetic studies when reactions are very rapid.

In such 'open' systems, the reactants are constantly renewed in such a manner that an apparently stationary state is established in the reactor: this composition is identical in all parts of the reaction mixture up to the effluent if stirring is sufficiently strong.

A variant of this procedure consists in using tubular continuous flow reactors, where the concentration is constant in each cross-section of tube but varies along its length.

While the account of kinetics in the first two chapters is sufficient for static procedures, it is insufficient if we are to consider dynamic procedures which give directly the rate of a chemical reaction. Accordingly, the rest of this chapter will cover the kinetics of tubular and continuous flow reactors, followed by a study of modern experimental methods for measuring the rate of fast reactions, completed in less than about one minute. The reader will find the description of the classical experimental procedures in the detailed texts listed at the end of the book.

A. *Continuous Flow Reactors*

The volume V of the reaction mixture is less than that of the reactor. The composition and temperature must be identical at all times and in all parts of the reactor; $\bar{n}_i^0$ and $\bar{n}_i$ are the molar fluxes (number of moles per unit time) of compound i entering and leaving the reactor.

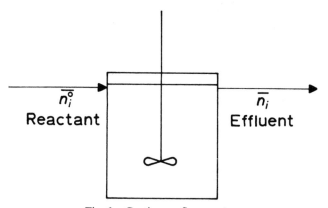

$\overline{n_i^0}$
Reactant

$\overline{n_i}$
Effluent

Fig. 6.   Continuous flow reactor.

The difference $\bar{n}_i - \bar{n}_i^0 = \pm v_i V \mathscr{V}$ and the rate measured directly is given by

$$\mathscr{V} = \pm \frac{\bar{n}_i - \bar{n}_i^0}{v_i V} \, .$$

If the reactant is a limiting reactant 1, we have

$$f = \frac{n_l^0 - n_l}{n_l^0} = \frac{\bar{n}_l^0 - \bar{n}_l}{\bar{n}_l^0}$$

and

$$\frac{\bar{n}_l^0 f}{V(-v_l)} = \mathscr{V}.$$

The rate of the reaction is then equal to the limiting number of moles of reactant feeding the reactor per unit of time and of reaction volume, multiplied by the fractional conversion (if $-v_l = 1$).

If P is a product present in the effluent and not in the incoming fluid, it follows that

$$-\frac{\bar{n}_l^0}{v_l} f = \frac{\bar{n}_p}{v_p}$$

so that

$$\frac{\bar{n}_p}{V} = v_p \mathscr{V}$$

and we may call the quantity $\bar{n}_p/V$ the space-time yield.

If $v_p = 1$, and if the reactor functions under ideal conditions (which is not always the case in practice), the rate of the reaction is equal to this yield, whether the volume of the reaction mixture remains constant or not.

Very often, when operating in a solvent, the change in volume is very small; in that case, if we can measure the concentration of limiting reactant C_l^0 and C_l, before and after passage through the reaction, the volume F of flux being the same for input and effluent, then

$$F(C_l^0 - C_l) = V(-v_l)\mathscr{V} \quad \text{or} \quad (-v_l)\mathscr{V} = \frac{C_l^0 - C_l}{V/F}.$$

For a product P this formula becomes

$$\frac{C_p}{V/F} = v_p \mathscr{V}.$$

The ratio V/F corresponds to the residence time of the mixture through the reactor, and its inverse to the speed of passage.

In fact, there is not a unique residence time but an ensemble of times following a distribution law: some particles stay only a short time in the reactor, others much longer, and V/F represents only the average residence time.

The use of continuous flow reactors is not only applicable to first- and second-order reactions, but above all is useful for complex processes, because it enables us to avoid the integration of differential equations of rate.

EXAMPLE 1.   Second-order reactions.

$A + B \rightarrow$ product.

If the initial concentrations of reactants before mixing are $2a$ and $2b$, and if x is the amount of product formed at time t, then

$$\frac{x}{t} = \mathscr{V} = k(a - x)(b - x)$$

with $v = 1$ and $t = V/F$ residence time.

EXAMPLE 2.   Interaction of products with reactants.

$$A + B \xrightarrow{k_1} C + D$$
$$A + C \underset{k_3}{\overset{k_2}{\rightleftarrows}} P \quad \Bigg\} \quad \text{with} \quad K = \frac{k_2}{k_3}.$$

At time t
$$\begin{cases} |A| = a - x - P \\ |B| = b - x \\ |C| = c + x - P. \end{cases}$$

For $t = 0$
$$\begin{cases} |A| = a \\ |B| = b \\ |C| = c \\ |D| = 0. \end{cases}$$

In a static reactor, the rate equation is

$$\frac{dx}{dt} = k(a - x - P)(b - x)$$

with

$$K = \frac{P}{(c + x - P)(a - x - P)} \rightarrow$$

leading to

$$2P = \left(\frac{1}{K} + C + a - \sqrt{\left(\frac{1}{K} + C + a\right)^2 - 4(c + x)(a - x)}\right).$$

The integration of the complete expression is difficult.
In a continuous flow reactor

$$\frac{x}{t} = k(a - x - P)(b - x).$$

After having substituted for P, an expression is obtained

where only k and K are unknown. It is necessary to carry out two experiments (varying a, b, or c) to obtain these.

EXAMPLE 3. Consecutive second-order reactions.

The basic hydrolysis of a diester can be examined by this procedure.

$$
\left\{
\begin{array}{ccccc}
OH^- & + RO_2C - CO_2R & \xrightarrow{k_1} & RO_2C - CO_2^- & + ROH \\
a - x - y & b - x & & x - y & \\
OH^- & + RO_2C - CO_2^- & \xrightarrow{k_2} & {}^-O_2C - CO_2^- & + ROH \\
a - x - y & x - y & & y &
\end{array}
\right\}
$$

The change in the concentration of the ion OH^- is followed as a function of time t, with a and b as initial concentrations of OH^- and diester.

Then

$$
\frac{x}{t} = k_1 |OH| \, |b - x| \qquad k_1 = \frac{x}{|b - x| \, |OH| \, t}
$$

$$
\frac{y}{t} = k_2 |OH| \, |x - y| \qquad k_2 = \frac{y}{|x - y| \, |OH| \, t}.
$$

Two experiments are necessary, since the two equations contain, as well as $OH = a - x - y$, four unknowns.

B. *Tubular Reactors*

The reactor is a cylindrical tube of constant cross-section completely filled with the reaction mixture which is forced through it by pistons.

The two reactants meet in a mixing chamber which can be more or less complicated according to the duration of the reaction; the simplest is a T-tube, the right-angled side-arm constituting the reactor.

The composition of the mixture changes from section to section, but remains constant in a given section if convection currents and diffusion are avoided. All elements of volume thus have the same residence time.

For a limiting reactant

$$\mathrm{d}n_l = v_l \, \mathrm{d}\mathrm{V}_r \mathscr{V} \quad \text{hence} \quad - n_l^0 \frac{\mathrm{d}f}{\mathrm{d}\mathrm{V}_r} = - v_l \mathscr{V}.$$

Here the volume of the reaction mixture is equal to the volume of the reactor $V = V_r$. If the change in volume of the reaction mixture is negligible, the rate in the reaction does not vary, and the residence time is V_r/F, with F equal to the flux entering the apparatus. In a liquid medium this condition is realized essentially, so that

$$\frac{\mathrm{d}C_l}{\mathrm{d}\left(\dfrac{\mathrm{V}_r}{\mathrm{F}}\right)} = \frac{\mathrm{d}C_l}{\mathrm{d}t} = v_l \mathscr{V}.$$

With a gaseous medium, this is the case only if the reaction is not accompanied by a change in the number of gaseous molecules: if $\Delta n > 0$, the rate of passage increases and the residence time is less than V_r/F.

Observation at a distance d_{cm} from the mixing chamber is effected by an appropriate physico-chemical technique.

At this point, the residence time has been

$$t = \frac{d_{cm}}{u_{cm\,s^{-1}}} = \frac{V_r}{F},$$

where u is the speed of the fluid.

By varying the speed of passage u, or by observing at a variable distance d, we can obtain directly the variation of concentration as a function of time.

This method is very useful for the study of fast reactions, but can be adapted to slower reactions by interposing an appreciable dead volume between the mixing chamber and the observation window.

3.3. Application to the Study of Fast Reactions

Actual chemical reactions are rarely simple, but usually take place by a sequence of elementary reactions, some of which can be very rapid. Some of the intermediates are accordingly very fleeting and very difficult to demonstrate. The complete elucidation of a reaction mechanism can often be realized only by the aid of special kinetic methods.

The duration of a chemical reaction, even in the case of a simple intramolecular rearrangement, cannot be less than the period of a physical process such as a molecular vibration (of the order of 10^{-12} s). Chemical processes have accordingly a greater duration (10^{-11} s at ordinary temperature), but this quantity is excessively small for ordinary kinetic methods, which scarcely allows us to work with times less than about one minute. The domain between 1 min and 10^{-11} s has been the object of numerous researches in the past thirty years, and while in current practice it is still difficult to work at less than a millisecond, various special methods have allowed us to study reactions with times of the order of a nanosecond, as shown in Table I.

TABLE I

	Interval
Continuous flow } Stopped flow }	1 min–10^{-3} s
Temperature jump	1 min–10^{-5} s 10^{-6}
Pressure jump	10^{-5}–10^{-8} s 10^{-9} ?
Potential jump *field jump*	10^{-4}–10^{-8} s
Ultrasonic	10^{-5}–10^{-9} s
Nuclear magnetic resonance	10^{-1}–10^{-6} s
Electron paramagnetic resonance	10^{-4}–10^{-9} s
Flash photolysis	1–10^{-6} s
Pulse Radiolysis	1–10^{-7} s 10^{-9}
Electrochemical methods	1–10^{-6} s

a – switche lasers 10^{-9} s. – 10^{-12}

The complete description of these methods is beyond the scope of this book, and we shall confine ourselves to giving some indication of their utilization and of the calculations necessary to exploit them; their fruitfulness in experimental kinetics is becoming ever more apparent.

3.4. Competition Methods

A. *Continuous Flow*

The methods of continuous flow in tubular reactors are

adaptable to a large number of cases and have allowed the classical methods to be applied to lifetimes down to a centisecond. They were introduced in 1923 by Hartridge and Roughton, but their full development has coincided with recent progress in electronics.

Principal. The two reactants are fed into the reactor by the pressure of electrically-operated pistons in automatic burettes or by the pressure of an inert gas. For very rapid reactions the efficiency of mixing must be very great, because the mixing time for the two fluids must be less than the period of the reaction; under optimal conditions it can go down to about a millisecond and the flow rate can attain 20 m s^{-1}.

The method for observing the reaction is most often spectrophotometric (although other physical methods have been used, such as the variation of pH for the reaction $CO_2 + \bar{O}H \rightarrow HCO_3^-$); when the stationary state is attained, it often becomes possible to obtain the absorption spectra of intermediates in the reaction. Additionally, by varying the rate of flow and the distance for observation from the mixing chamber it is possible to calculate the rates of the different steps.

The method would be most versatile if it did not require large quantities of reactants (a minimum of 20 cm^3) for each experiment.

B. *Electrochemical Methods*

In these methods, and more particularly in polarography,

we measure the intensity of an electric current, which results in a complex manner from the competition between the processes of diffusion and of reaction; contrary to other methods, where concentration changes are observed directly, we are concerned here with an indirect procedure. This disadvantage is compensated by the advanced state of development, theoretical and experimental, of the technique, due to non-kinetic applications; the equipment has been commercial for a long time and is not expensive.

In fact, for reactions having a duration of more than 30 s, polarography can be used in the classical manner to analyse rapidly the reaction mixture, by keeping the polarization potential constant; this microanalytical method requires only a small quantity of reaction mixture, but it requires an excess of foreign electrolyte to be included in the mixture, and this may perturb the mechanism of an ionic reaction.

For very rapid reactions it is necessary to use indirect procedures (kinetic or catalytic currents) which are relatively easy to carry out, but which give results difficult to interpret in terms of rate coefficients, so that the values obtained are not always in agreement with those obtained by other methods.

Chronopotentiometry and electrolysis at a constant current are more recent, but the results already obtained seem promising. Similarly, the use of a rotating disc electrode in place of the dropping mercury electrode allows the surface to be continually renewed, so that the diffusion layer has a thickness independent of time (but a function of the angular velocity of the disc). From the value of the current in this layer, when it is limited by the rate of homogeneous reaction, we can obtain the value of the rate coefficient.

The stationary electric field method of Eigen enables rate coefficients of ionic reactions to be measured at very low concentrations (because the solution must have a very low conductivity); it has led to the measurement of the rate of neutralization of H^+ by OH^- ions in ice, and can be applied to non-aqueous solvents and even to semi-conductors.

Between two electrodes very close together (0.01 cm) a potential difference of 1000 V is established. All the ions migrate towards the corresponding electrodes, so that the saturation current can come only from the formation of new ions.

EXAMPLE: In very pure ice, the ions can come only from the dissociation of water molecules, thus

$$H_2O \underset{k_2}{\overset{k_1}{\rightleftarrows}} H^+ + OH^-$$

and the intensity of the residual current is $i_k = z\mathscr{F}Alk_1 \times (H_2O)$ where A is the surface of the electrodes, l is the distance between the electrodes, and $z\mathscr{F}$ is the charge corresponding to 1 mole of H_2O.

Another example of the application of electrochemistry to follow fast reactions is the concentrostatic method of J. E. Dubois.

By coulometric regeneration, the concentration of bromine is keep constant but very low for a second-order reaction (bromination of olefins), so that the rate of bromination remains sufficiently low to be measured: this rate is proportional to the intensity of the current.

The method is applicable only to bimolecular reactions, but allows constants of 10^3 l mole^{-1} s^{-1} to be measured, and has wide applicability.

C. *Photostationary Methods*

An external influence can produce a stationary state (photochemical equilibrium) clearly different from that which exists in its absence, but requires a continual influx of energy (visible or ultraviolet light) which is dissipated in the reaction mixture.

This technique has been used to produce free radicals and to study their recombination, and also to examine the fluorescence produced in competitive processes between 'excited' molecules.

The first type of study requires us to know the quantum yield and the absolute concentration of all chemical species present, while the second requires only relative concentrations, since the reactions are second-order.

Recombination of free radicals. Symmetrical molecules (Cl_2) dissociate under the influence of light to give free atoms, which recombine to give the initial molecule.

A stationary state is established when the rates of the two processes become identical

$$\left.\begin{array}{l} X_2 + h\nu \rightarrow 2X^{\cdot} \\ X^{\cdot} + X^{\cdot} \rightarrow X_2 \end{array}\right\}$$

so that

$$k = \frac{\Phi q}{|X|^2}$$

where k is the rate coefficient in $1 \text{ mole}^{-1} \text{ s}^{-1}$; Φ is the quantum yield, q is the rate of light absorption in einstein 1^{-1} s^{-1}, $|X|$ is the stationary concentration of free atoms.

It is possible to obtain q by spectrometry of actinometry; and the stationary concentration $|X|$ by means of augmenting or diminishing its concentration or that of X_2.

In the gaseous phase, the technique is applied in the following manner: the gas is illuminated strongly in one direction, while a secondary beam of light directed perpendicularly to the first is used to measure spectrophotometrically the concentration of molecules. The halogens have been studied in this way, because the rate of their recombination is relatively slow (the process is in reality trimolecular).

In the liquid phase, the diminution in the concentration of molecules becomes negligible, and it is easier to measure directly the concentration of atoms or radicals by electron spin resonance, although it is difficult to calibrate the apparatus to obtain absolute concentrations (which are always less than $10^{-7} \text{ mole } 1^{-1}$).

The rate of recombination of bromine atoms is 2.4×10^9 l^2 mole^{-2} s^{-1}, in reasonably good agreement with the value of 3.6×10^9 found by using flash photolysis, and the photostationary method can thus be used to obtain reaction rates much greater than those obtained by other competitive methods.

Use of fluorescence. The absorption of light can yield 'excited' molecules, which return to their ground state by many routes, of which one consists of the emission of photons, generally having a wavelength longer than that of the exciting radiation. The intensity of the fluorescence is proportional to the stationary concentration of excited molecules, formed by constant illumination.

The excited molecules are more reactive and hence their reactions more rapid than those of ordinary molecules; fluorescence measurements give information directly on the competition between the deactivation of the excited molecules, whose average life is of the order of 10^{-8} s, and their reactions, which can be very rapid.

Numerous reactions of dimerization (of pyrene, anthracene) of the type

$$A + h\nu \rightarrow A^* \begin{cases} A^* + A \rightarrow A_2 \\ A^* \rightarrow A + h\nu' \end{cases}$$

of complex formation, of protonation (acridine), etc., have been studied and found to have activation energies less than 5 kcal mole^{-1}

3.5. Perturbation or Relaxation Methods

A. *Stopped Flow*

The method of continuous flow (Ia) has been modified by the sudden stoppage of the flow, and by the continuous observation of the extent of the reaction at a very small distance from the mixing chamber. The advantages are very great, because:

(a) The volume of reactants is reduced to 10^{-1} cm^3 per experiment;

(b) The observation is independent of the volumetric flux (and its variations) and of hydraulic distortions;

(c) If the procedure for registering the change is sufficiently rapid, observation can be made from 10^2 s to a few milliseconds after mixing.

In fact, registering requires the use of an oscilloscope coupled to the movement of the pistons, followed by photographing of the screen.

The method is a little less sensitive than the continuous flow method, and allows a spectrum to be obtained only point by point, so that it is less well suited to measuring the rates of fast reactions at temperatures between $-50°$ and $+40°$C.

B. The limitation of the preceding method (to reaction times greater than about 2 ms) is due essentially to the time necessary to achieve complete mixing of the two reactants isothermally. Towards 1950, Eigen succeeded in avoiding this difficulty by effecting a sudden perturbation in a reaction mixture in an equilibrium or stationary state, and hence already mixed; this perturbation causes a chemical displacement, and the velocity of return to equilibrium can be studied.

The perturbation can be simple or produced periodically, and in the latter case, according to the frequency and the rate of the reaction, the chemical displacement will follow exactly or with a certain time lag the perturbation.

Among the three perturbations considered below, the first seems to be the most generally applicable.

Temperature jump. Chemical equilibrium depends on temperature according to the isochore

$$\frac{\partial \log Kp}{\partial T} = \frac{\Delta H_0}{RT^2}$$

so that a change in temperature causes a change in concentration of constituents.

However, the change in temperature must be produced very rapidly. This can be done by the discharge of a condenser across a mixture (made sufficiently conductive, if necessary by the introduction of a neutral salt) contained in a cell with two electrodes. The temperature rise increases exponentially according to $e^{2/RC}$ where R is the electrical resistance of the mixture and C the capacity of the condenser: it can reach $6°$ in 1 cm^3 in 10^{-7} s.

The registering of the changes in concentration must

then be very rapid, so that the apparatus for this technique is often coupled with that for stopped flow.

Pressure jump. The reaction mixture, submitted to pressures of 10 to 50 atm, is suddenly decompressed by the rupture of a diaphragm. The speed of the perturbation is limited only by the speed of sound.

Detection methods must be very sensitive, because pressure variation is less effective chemically than temperature variation, and in fact has been limited to differential measurements of conductivity.

Variation of an electric field. This variation is of interest in the case of ionic reactions, but requires a field of the order of 100 kV cm^{-1}. The resistivity of the bath must now be very high to limit heating by the Joule effect, so that it is necessary to work with very dilute solutions.

These three methods are based on the displacement of a chemical equilibrium, and their practical application depends essentially on the kinetics of establishing these equilibria (Chapter 1). But because the displacement obtained is relatively small, the rate equations become simplified, which allows for a rapid exploitation of the results.

For a reversible reaction of the type $A + B \underset{k_2}{\overset{k_1}{\leftrightarrows}} C$, the rate equation is

$$\frac{dC}{dt} = k_1 AB - k_2 C.$$

At equilibrium $t \rightarrow \infty$, $A = A_\infty$, $B = B_\infty$ and $C = C_\infty$.

Close to equilibrium, one can write

$$A = A_\infty + \Delta A \quad B = B_\infty + \Delta B \quad C = C_\infty + \Delta C$$

where

$$\Delta A = \Delta B = - \Delta C.$$

Introducing these into the rate equation, we obtain

$$\frac{d(\Delta C)}{dt} = [k_1 (A_\infty + B_\infty) + k_2] \, \Delta C \\ - k_1 A_\infty B_\infty + k_2 C_\infty - k_1 (\Delta C)^2.$$

Close to equilibrium $(\Delta C)^2$, infinitely small second order, can be ignored

$$\frac{d(\Delta C)}{dt} = - [k_1 (A_\infty + B_\infty) + k_2] \Delta C = - \frac{\Delta C}{\mathcal{H}}$$

where $\mathcal{H}$, the relaxation time, is equal to

$$\mathcal{H} = \frac{1}{k_1 (A_\infty + B_\infty) + k_2}$$

so that

$$\Delta C = \Delta C_0 \, e^{-t/\mathcal{H}} \quad \text{with} \quad \Delta C_0 \quad \text{at} \quad t = 0.$$

The second-order approximation above allows us to subsume all rate equations (close to equilibrium) to first-order equations. Furthermore, the measurement of the relaxation time $\mathcal{H}$ allows us to calculate k_1 and k_2 if the equilibrium constant $K = k_1/k_2$ is known.

Generalization. In the case of many coupled steps in an actual mechanism, if n is the number of independent variables there are n rate equations, which close to equilibrium all become first-order equations of the form:

$$-\frac{d\,(\Delta C_i)}{dt} = \sum_{k=1}^{k=n} a_{ik}\Delta C_k$$

where the coefficients a_i are functions of the rate coefficients and the concentrations at equilibrium. The solution for such a system requires the use of matrix calculations, and the integrated equations all appear as sums of exponentials (first order). For each intermediate one can define a relaxation time

$$\frac{1}{\lambda_j} = \mathscr{H}_j$$

a function of all the equilibrium constants and rate equations, but which does not correspond to a determinate elementary step.

One obtains in this way a relaxation spectrum whose study as a function of concentration can yield all the rate constants.

EXAMPLE:

$$A + B \underset{k_{-1}}{\overset{k_1}{\rightleftarrows}} C \underset{k_{-2}}{\overset{k_2}{\rightleftarrows}} D$$

$$-\frac{dA}{dt} = k_1 AB - k_{-1}C$$

$$= k_1\,(A_\infty + B_\infty)\,\Delta A - k_{-1}\Delta C - k_2\Delta C + k_{-2}\Delta D$$

$$-\frac{dD}{dt} = -k_2 C + k_{-2} D = -k_2 \Delta C + k_{-2} \Delta D.$$

At constant volume the sum $\Delta A + \Delta C + \Delta D$ is by definition zero, so that

$$-\frac{d(\Delta A)}{dt} = \underbrace{\left[k_1 (A_\infty + B_\infty) + k_{-1}\right]}_{a_{11}} \Delta A + \underbrace{k_{-1}}_{a_{12}} \Delta D$$

$$-\frac{d(\Delta D)}{dt} = \underset{a_{21}}{k_2} \Delta A + \underset{a_{22}}{(k_{-2} + k_2)} \Delta D.$$

The relaxation times are obtained by calculating the determinant:

$$\left\{ \begin{matrix} a_{11} - \dfrac{1}{\mathscr{H}} & a_{12} \\[2ex] a_{21} & a_{22} - \dfrac{1}{\mathscr{H}} \end{matrix} \right\} = 0.$$

hence

$$\frac{1}{\mathscr{H}} = \frac{(a_{11} + a_{22}) \pm \sqrt{(a_{11} + a_{22})^2 + 4(a_{12} a_{21} - a_{11} a_{22})}}{2}$$

In the case when the monomolecular step is slower than the bimolecular,

$$k(A_\infty + B_\infty) + k_{-1} \gg k_{-2} + k_2$$

then

$$\frac{1}{\mathscr{H}_1} = k_1 (A_\infty + B_\infty) + k_{-1}$$

and

$$\frac{1}{\mathscr{H}_2} = k_{-2} + \frac{k_2}{1 + k_{-1}/k_1\,(A_\infty + B_\infty)}$$

whence

$$\frac{t}{\mathscr{H}_1} \gg \frac{1}{\mathscr{H}_2}.$$

These relaxation times are independent of the thermo-dynamic perturbation (although that measured by temperature jump should be isothermal while those measured by ultrasonics are adiabatic) and correspond to actual chemical processes.

Relaxation methods are methods of direct measurement analogous to classical methods and can be applied to all equilibrium states (or stationary states): such a state may imply a reaction intermediate which itself is in equilibrium, but the system however is 'open' unlike a closed equilibrium.

According to the perturbing force, these methods can be used to follow the decay in concentration as a function of time, or to study the frequency of the response if the impulse is repeated periodically. The latter case occurs with ultrasonic perturbation and, according to the relative value of the relaxation time of the system and the period of the external parameter, the displacement of the equilibrium can be rigorously stationary $(1/\omega \ll \mathscr{H})$, modulated with a lower frequency and amplitude $(1/\omega \# \mathscr{H})$, or follow the modulation of the perturbating force $(1/\omega \gg \mathscr{H})$.

These three methods are particularly suited to the study of complex systems, because the different steps appear directly on the time axis; the interpretation of the data is easier than that for competitive methods. In particular,

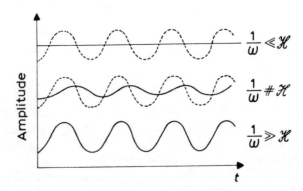

Fig. 7.   Amplitude and frequency of perturbing force.

these techniques have allowed Eigen to study protolytic reactions – protonations and deprotonations of normal acids and bases: diffusion – controlled processes which go in two stages (diffusion of the solvated proton or OH^- towards the base or acid, followed by rapid transfer of a proton across a complex formed by a hydrogen bond). The rates are very great, of the order of 10^{-10} M^{-1} s^{-1} (bimolecular), but are much slower for pseudo-acids (C—H), where the resultant ions are stabilized by an altered electronic structure. The rate of protonation of a C^- is lower by a factor of 10^3 than that of an O^-, and it is the same for deprotonation, which now no longer is diffusion-controlled.

However, relaxation methods require costly and delicate apparatus, especially when it is necessary to go to times of less than a microsecond, and cannot be used to examine non-macroscopic reactions such as the exchange of a proton between two identical molecules.

C. *Flash Photolysis and Pulsed Radiolysis*

These two methods bring into play much more important perturbations than the preceding methods, so that their interpretation is difficult, because the rate equations no longer reduce to first-order equations.

Flash photolysis, although less specific than the photo-stationary method, which is better adapted to the determination of primary processes, allows us above all to prepare unstable intermediates in unusually high concentration, in gases, liquids, and solids.

The flashes of visible or ultraviolet light can have a very short duration but can be very intense, in order to produce a noticeable change in concentration in the reaction mixture in about a microsecond.

The necessary equipment is easily available and the experimentation does not pose any particular problem (Figure 8); detection is in general by spectrophotometry, and the measurement of the intensity of the flashes can be made by actinometers of ferrioxalate or uranyl oxalate.

The only difficulty lies in the fact that neither the concentrations, nor the absorption coefficients of the reaction intermediates are known, so that except for first order reactions, rate constants can only be calculated indirectly.

This method has allowed Norrish and Porter to study a large number of free radicals, either simple such as $CH_2^{\cdot}$, or $NCS^{\cdot}$ or more complex such as aromatic radicals. At the same time the reaction of chlorine and oxygen has been studied, with that of the intermediate ClO.

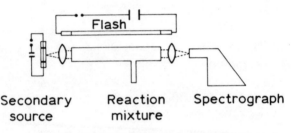

Secondary   Reaction   Spectrograph
source   mixture

Fig. 8.   Sketch of flash-protolysis apparatus.

3.6. Intermediate Methods

A. *Nuclear Magnetic Resonance*

This method is particularly well adapted to simple mechanisms and to microscopic reactions (exchange), such as $NH_3 + NH_2^- \rightarrow NH_2^- + NH_3$, in the system $NH_3 - NH_2Na$.

It is necessary that the system have a resonance spectrum in which the peaks can be identified. When the rate of exchange is altered, by change of temperature, increased amount of catalyst, etc., different peaks of the spectrum coalesce. Since the displacement of the frequency can vary from 10 to 10^3 cycles s^{-1}, the method is directly applicable to reactions having mean times between 10^{-1} and 10^{-3} s. Furthermore, by studying the width of bands as a function of concentrations, one can extend the range of applicability to a microsecond.

B. *Paramagnetic Electron Resonance*

This applies particularly to the identification of free radicals, but the spectra obtained are much more complex, which has limited kinetic applications.

CHAPTER 4

REACTIONS IN THE GAS PHASE

Chemical reactions between stable chemical entities are rarely simple, and in passing from reactants to final products complex rearrangements of chemical bonds are required, which lead to intermediate particles or *reactive centres*. In general, a reaction usually involves a sequence of elementary changes, even though the ensemble can be still described by a single kinetic parameter (the extent of reaction) and can be stoichiometrically simple.

The form of the rate expression for an actual reaction depends on the rates of the elementary steps, but it suffices for one step to be noticeably slower than the others for it to impose its rate on the system (to be the rate-determining step).

In general, the experimental study of reactions in the gas phase is rather difficult, but the theoretical interpretation is easier than in solution because of the simpler environment of the molecules.

Two principle types of reaction are found in gas phase reactions:

Open sequences: the different elementary steps always take place in the same order from reactant to product.

Closed sequences: a reactive centre is regenerated in one step and enters again into the reaction, so that a certain number of steps recur in cyclical fashion (propagation steps) and a single active centre can lead to a great number of molecules of product.

These two groups can be described in a simple way by using a method of calculation based on the principle of a quasi-stationary state.

4.1. Steady State Principal

The treatment of a simple system $A \xrightarrow{k_1} B \xrightarrow{k_2} C$ already given in Chapter 1 indicates that in the case where B is a highly reactive intermediate, i.e. $k_2 \gg k_1$ or $k_1/k_2 \to 0$, the expressions for (A), (B) and (C) are simplified and become

$$|A| = ae^{-k_1 t} \quad |B| = a \frac{k_1}{k_2} e^{-k_1 t} = \frac{k_1}{k_2} |A|$$

$$|C| = a(1 - e^{-k_1 t}).$$

The point of inflexion of the curve representative of C is very close to the origin, as is also the maximum in the curve giving the amount of B against time. The three equations above correspond to

$$\frac{d|A|}{dt} = -k_1 |A| \quad k_1 |A| - k_2 |B| = 0$$

$$\frac{d|C|}{dt} = k_2 |B|.$$

The middle equation signifies that $d|B|/dt = 0$ or that the rate of appearance of the active centres B is equal to their rate of disappearance at an instant t (stationary state). This equation, however, cannot be integrated because the concentration of B in reality is never constant, because it is dependent on that of A which decreases exponentially. In the case of a stationary state $(k_2 \gg k_1)$, this concentration of B is very small compared to that of A and of C (reactant and product), and from the equality of the rates of appearance and of disappearance comes the fact that the actual reaction, even although complex, can be described by a single term for extent of reaction.

Remark 1: The stationary state is reached in the time t_r or relaxation time, necessary for the maximum concentration of B to be reached, which is equal to the average lifetime of an active centre, and which yields $t_r = 1/k_2$.

Remark 2: The approximation given by the steady state principle holds very well in most cases encountered; exceptions are so rare that in general it can be used to describe actual reactions. Thus, for the open sequence

$$A \overset{k_1}{\to} B \to C \to \to R \to S,$$

where s is the stable end product, if $k_1 \ll k_i$, the first step is rate-determining for the sequence and constitutes a 'bottle-neck'. B, C, $---$ R are fugitive intermediates of very low concentrations, which may be calculated by application of the steady-state principle to each step. It follows that

$$k_1 |A| = k_2 |B| = k_i C_i$$

and that the ensemble is equivalent kinetically to $A \overset{k_1}{\to} S$

with

$$|S| = a\left(1 - e^{-k_1 t}\right)$$

$$|i| = \frac{k_1}{k_i} a\, e^{-k_1 t}.$$

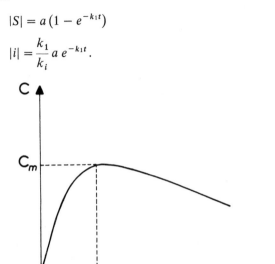

Fig. 9. Quasi-stationary state.

4.2. Reaction Intermediates or Active Centres

These intermediates are chemical entities having a reactivity depending on how close the energy minimum is to the energy maximum (activation barrier) they must surmount (Figure 4), and from this fact must often possess a special structure.

In the gas phase the active centres encountered most

often are either molecules having an atom in an abnormal valency state, or free atoms or radicals.

EXAMPLE:

Molecules → dichlorocarbene CCl_2 (divalent carbon);
Free radicals → methyl $CH_3^{\bullet}$.

The great reactivity of these intermediates gives them very short lives at ordinary temperatures, and makes it very difficult to demonstrate their presence. Besides the vapours of the monoatomic metals (Na), the first free atoms to be known were hydrogen and the halogens, which form at high temperatures by dissociation of their molecules

$$H_2 \,(+\,M) \leftrightarrows 2H^{\bullet}\,(+\,M) \quad \Delta H = +\,102 \text{ kcal mole}^{-1}$$

T^0...	2000	2500	3000	3500	4000	
% disso-ciation...	0.08	1.26	7.9	28.6	62.3	$p = 1$ bar

but the two methods used for studying their physico-chemical properties are the following:

Electric discharge (Wood): under 2000 V, an electric discharge is caused in a long tube (1 m) containing the gas at low pressure. The two extremities of the tube have optical surfaces so that the gas can be observed. If the internal surface of the tube is covered by syrupy phosphoric acid, the heterogeneous process for recombination of the atoms is inhibited, and the free atoms have a sufficient life-time (average life of $H^{\bullet}$ atoms is 0.33 s under a pressure of 0.5 torr) to be pumped into another container.

The recombination of free atoms takes place by a trimolecular process, characterized by a very low energy of activation, or on contact with the wall of the container by a heterogeneous mechanism (application: atomic blow-pipe of Langmuir).

In certain cases, the molecules obtained are in an 'excited' state and give off a fluorescence in becoming deactivated:

$$N^{\bullet} + N^{\bullet} + N_2 \rightarrow N_2 + N_2^*$$
$$\searrow^{h\nu} \text{ green}$$
$$N_2$$

From the intensity of the fluorescence one can deduce the rate of the reaction.

According to the principle of microreversibility, the dissociation of diatomic molecules A_2 requires an enthalpy of activation of the order of magnitude of the enthalpy of dissociation, and can be brought about only at high temperatures.

TABLE II

	E	ΔH	θ_c^0
I_2...	30	34.5	1000–1700
Br_2...	40	44.6	1200–2100
Cl_2...	49	56.6	1700–2600
H_2...	95	102.7	2600–4000

Photochemistry. Molecular hydrogen containing traces of mercury vapour (photosensitizer) is irradiated by UV radiation ($\lambda = 253.7$ nm). The gas can then bring about hydrogenation reactions at ordinary temperatures without catalyst. The mechanism is the following

$$\begin{cases} Hg + h\nu \underset{\text{photon}}{\rightarrow} Hg^*_{\text{excited}} \\ Hg^* + H_2 \rightarrow HgH + H^{\bullet}. \end{cases}$$

Free radicals. Evidence for the existence of free radicals came in 1900 from Gomberg, who showed that hexa-phenylethane in benzene solution dissociates spontaneously into a free 'triphenyl methyl' radical

$$\underset{\underset{80\% \text{ colourless}}{}}{(C_6H_5)_3C - C(C_6H_5)_3} \rightleftarrows \underset{\underset{20\% \text{ yellow}}{}}{2(C_6H_5)_3C}.$$

The solutions have abnormal chemical properties, as shown by the absorption of molecular oxygen in the cold

$$2(C_6H_5)_3C^{\bullet} + O_2 \rightarrow \underset{\text{peroxide}}{(C_6H_5)_3C - O - O - C(C_6H_5)_3}.$$

However, it is only since 1925–1930 that the study of free radicals developed significantly with the experiments of Paneth.

A vacuum of 1 to 2 torr is maintained in a tube by a high-capacity pump; a stream of gas (N_2 or H_2) is passed along the tube at velocities around 14 m s^{-1} and picks up the lead tetramethyl from a wash-flask. When the tube is heated at B a lead mirror is formed by the reaction

$$Pb(CH_3)_4 \rightarrow \underset{\downarrow}{Pb} + 4CH_3^{\bullet}.$$

When the tube is heated at A, a new mirror forms here, while the first mirror disappears if the distance AB is less

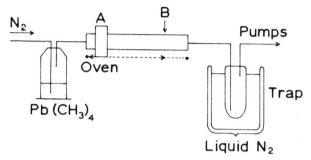

Fig. 10. Apparatus of Paneth.

than a critical length depending on the temperature and the velocity of the gas stream; in effect, for the mirror B to disappear it must be reached by the radicals formed at A, whose short lifetime can thus be determined. Since that time mirrors have often been used to characterize free radicals: thus Rice used tellurium mirrors to distinguish between methyl and methylene. However, the characterization of free radicals has resulted from the development of new physico-chemical methods, particularly spectroscopic procedures.

Absorption spectra $\rightarrow (C_6H_5)_3C^{\cdot}$.
Emission spectra $\rightarrow OH^{\cdot}, CH^{\cdot}, CN^{\cdot}, NH^{\cdot}, SH^{\cdot}$

Mass spectra $\rightarrow$ method of choice, also utilizable for following the course of pyrolyses or combustions.

Measurements of magnetic susceptibility have been used because of the paramagnetism due to the unpaired electron of free radicals. The study of their properties has often taken place at low temperature by freezing in liquid helium at 4K, inclusion in plastic materials, etc. Their lifetime at ordinary temperatures in general is short, and the recombination of polyatomic particles takes place by bimolecular processes $2CH_3^{\cdot} \rightarrow C_2H_6$. Certain free radicals show a greater stability, such as di-t-butylnitroxide $[(CH_3)_3C]_2NO$, which occurs as a red liquid, capable of being distilled at 75°C under 35 torr.

The methods of flash photolysis allow detectible concentrations of free radicals to be observed, by exposing the reaction mixture to a flash of great intensity and then registering the spectrum immediately afterwards.

The production of a shock wave in a tube containing a gaseous mixture causes adiabatic heating of successive layers of gas, and can be of value for free radical studies.

4.3. Chain Reactions

Reactions in closed sequence may be differentiated into *unbranched chain reactions* if, in the cyclic reactions, one active centre gives rise to a single chain carrier, and *ramified (branched) chain reactions*, if it gives rise to two or more active centres in one of its steps.

In general, unbranched chain reactions can be described quantitatively, if the mechanism of bond rupture is simple, by application of the steady-state principle to active centres.

On the other hand, ramified chain reactions are very difficult to describe, since they terminate in explosions,

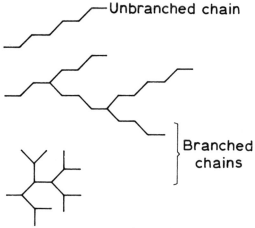

Fig. 11. Chain reactions.

but their study has developed strongly with the work of Hinshelwood and Semenov since 1955.

A. *Unbranched Chains*

Rice and Herzfeld studied many pyrolyses (heating in the absence of air) of organic substances, and have deduced an unbranched-chain mechanism for all these reactions.

The first step consists in the formation of an active centre, either by thermal dissociation at high temperature, by intervention of a catalyst, or by photochemical dissociation: this initiation reaction is the slowest.

The following steps propagate the chain, with two kinds of chain carriers: bimolecular free radicals R_1, and monomolecular free radicals R_2.

Initiation   M $\overset{1}{\to}$ R_1 + Prod.

Propagation R_1 + M $\overset{2}{\to}$ R_2 + Prod. $\Big\}$

Propagation R_2 $\overset{3}{\to}$ R_1 + Prod.

The rate expression depends essentially on the termination step: if one of the types 4a, 4b or 4c predominates, the form of the expression is relatively simple, and the reaction has an apparent order.

Termination	Order
$R_1 + R_1 \overset{4a}{\to}$ Product $\to \frac{3}{2}$	
$R_2 + R_2 \overset{4b}{\to}$ Product $\to \frac{1}{2}$	
$R_1 + R_2 \overset{4c}{\to}$ Product $\to 1$.	

Activation energies are from 30 to 60 kcal for initiation, less than 10 kcal for propagation steps, and very low for termination reactions, but the rates of the termination reactions are very small because the concentrations of free radicals are very small: the products obtained in weighable amounts arise from the propagation steps, and only traces (often difficult to isolate) of termination products are obtained, so that the stoichiometry of the reaction remains simple.

Chains can be short (10 to 20 steps) or long (to 10^6 steps) and their length represents the number of cycles which an average radical can effect after its formation and before the chain is terminated

$$1 = \frac{\text{disappearance of M}}{\text{formation of } R_1}$$

$$= \frac{k_1 |M| + k_2 |R_1| |M|}{k_1 |M|} \# \frac{k_2}{k_1} |R_1|.$$

EXAMPLE 1. Pyrolysis of acetaldehyde at about 350°C. Overall reaction

$$CH_3CHO \rightarrow CO + CH_4.$$

Mechanism

(1) $CH_3CHO \rightarrow CH_3^{\cdot} + CHO^{\cdot}$ thermal initiation.
$\quad\quad\quad C_1 \quad\quad\quad C_2 \quad\quad C_3$

(2) $CHO \rightarrow CO + H^{\cdot}$
$\quad\quad\quad\quad\quad\quad C_4$

(3) $H^{\cdot} + CH_3CHO \rightarrow H_2 + CH_3CO^{\cdot}$
$\quad\quad\quad\quad\quad\quad\quad\quad\quad\quad C_5$

(4) $CH_3CO^{\cdot} \rightarrow CH_3^{\cdot} + CO$

(5) $CH_3^{\cdot} + CH_3CHO \rightarrow CH_4 + CH_3CO^{\cdot}$

} propagation

(6a) $2CH_3^{\cdot} \rightarrow C_2H_6$

(6b) $CH_3^{\cdot} + CH_3CO^{\cdot} \rightarrow CH_3-CO-CH_3$

(6c) $2CH_3CO^{\cdot} \rightarrow CH_3COCOCH_3$

} termination

The predominant chain termination is of type (6a), but traces of acetone and biacetyl (identifiable spectroscopically) are also produced; kinetically (6b) and (6c) can be neglected. The rate of disappearence of CH_3CHO is

$$-\frac{dC_1}{dt} = k_1 C_1 + k_3 C_1 C_4 + k_5 C_2 C_1 .$$

The concentration of active centres can be calculated from the steady state principle:

$$
\begin{aligned}
CH^{\bullet} \quad &\rightarrow k_1 C_1 = k_2 C_3 \\
H^{\bullet} \quad &\rightarrow k_2 C_3 = k_3 C_1 C_4 \\
CH_3 \quad &\rightarrow k_1 C_1 + k_4 C_5 = k_5 C_1 C_2 + k_6 C_2^2 \\
CH_3CO^{\bullet} &\rightarrow k_3 C_1 C_4 + k_5 C_1 C_2 = k_4 C_5 .
\end{aligned}
$$

The $CH_3^{\bullet}$ free radicals are much the most numerous with a concentration

$$C_2 = \left(\frac{2k_1 C_1}{k_6}\right)^{1/2}$$

which gives

$$-\frac{dC_1}{dt} = 2k_1 C_1 + \left(\frac{2k_1}{k_6}\right)^{1/2} k_5 C_1^{3/2} .$$

Since k_1 (initiation) $\ll k_i$, the first term of this expression is often negligible in comparison with the second, so that this decomposition is found to be fairly well represented by the simple expression

$$-\frac{dC_1}{dt} = k' C_1^{3/2}$$

having an overall activation energy

$$E = E_5 + \tfrac{1}{2}(E_1 - E_6)$$

with

$E_1 = 76$
$E_5 = 10$
$E_4 = 18$
$E_6 \neq 0$.

In fact, with variation of experimental conditions the reaction can be made to pass from first to second order.

EXAMPLE 2. Decomposition of ethyl ether.

In the absence of oxygen, the decomposition takes place at a temperature of 300–400°C by the following chain mechanism, passing through the stage of 'acetaldehyde'

$(C_2H_5)_2O \rightarrow CH_3^{\cdot} + {}^{\cdot}CH_2OC_2H_5$
$CH_3^{\cdot} + C_2H_5OC_2H_5 \rightarrow C_2H_6 + {}^{\cdot}CH_2OC_2H_5$
${}^{\cdot}CH_2OC_2H_5 \rightarrow CH_3^{\cdot} + CH_3CHO$
$CH_3^{\cdot} + {}^{\cdot}CH_2OC_2H_5 \rightarrow C_2H_5OC_2H_5$.

The overall reaction is first order with an average chain length of 4.4.

4.4. Case of the Synthesis of Hydrogen Halides

The synthesis of the hydrohalic acids $X_2 + H_2 \leftrightarrows 2HX$ is a very interesting case of kinetics in the gas phase, to the extent that on going from iodine to fluorine the mechanism changes and becomes more complex, all the time being capable of .logical interpretation through different elementary steps.

A. *Hydriodic Acid*

$$H_2 + I_2 \leftrightarrows 2IH$$

The synthesis and the dissociation each follow a second-order law (even if the establishment of equilibrium follows a complex kinetic law: Chapter 1), and for many years these reactions were considered simple and bimolecular. However, although the steric factors lay between 0.3 and 1 (so that they constituted one of the best experimental verifications of elementary collision theory) the temperature coefficients were relatively too high.

In 1967, Sullivan showed that the mechanism could be more complex and proposed the scheme

$$I_2 \rightarrow 2I^{\cdot} \quad \text{with} \quad K = \frac{|I^{\cdot}|^2}{|I_2|}$$

$$2I^{\cdot} + H_2 \rightarrow 2HI$$

so

$$\frac{d|HI|}{dt} = k|I^{\cdot}|^2 |H_2| = kK|I_2| |H_2|.$$

The bimolecular step will be the slowest, and the general form of the rate equation is still second-order.

Kinetics does not enable us to differentiate directly between the two mechanisms, but the introduction of free iodine atoms (prepared by photochemical dissociation at

578 nm) into a mixture of reactants, at a temperature at which the thermal process is negligible, initiates the reaction. The enthalpies of activation are as follows (see Figure 12).

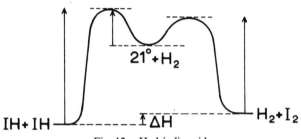

Fig. 12. Hydriodic acid.

which leads to

$$\Delta H = 43.8 - (35.5 + 5.3) = 3 \text{ kcal mole}^{-1}$$

for the reaction $2IH \rightarrow I_2 + H_2$.

The rate-determining step could be of the type $I^{\cdot} + H_2$, but the minimum activation energy would then be 50 kcal mole^{-1}, while the experimental value is 40.8 kcal mole^{-1}. This is in favour of the trimolecular step as rate-determining.

B. *Synthesis of Hydrobromic Acid*

$$Br_2 + H_2 \rightleftarrows 2BrH.$$

The first quantitative studies were done in 1906 by Bodenstein and Lindt and gave the following empirical expression, which fits well the numerical data at about 200–300°C

$$\frac{d|HBr|}{dt} = \frac{k|H_2||Br_2|^{1/2}}{m + \dfrac{|HBr|}{|Br_2|}}$$

where k and m are constants at a given temperature (m varies little between 25° and 300°, and is about 0.12).

The product HBr gradually inhibits the reaction, and the process does not have an order invariant with time; on the other hand, it has an initial overall order of $\frac{3}{2}$

$$\left[\frac{d|HBr|}{dt}\right]_0 \# k'|H_2||Br_2|^{1/2}.$$

A plausible mechanism was proposed towards 1920 and now seems to be well established

$$\begin{array}{lll}
Br_2 + M \xrightarrow{k_1} 2Br^{\cdot} + M & \Delta H_0 = 46 \text{ kcal mole}^{-1} \\
Br^{\cdot} + H_2 \xrightarrow{k_2} HBr + H^{\cdot} & \Delta H_0 = 16.6 \text{ kcal mole}^{-1} \\
H^{\cdot} + Br_2 \xrightarrow{k_3} HBr + Br^{\cdot} & \Delta H_0 = -41 \text{ kcal mole}^{-1} \\
H^{\cdot} + HBr \xrightarrow{k_4} H_2 + Br^{\cdot} & \\
2Br^{\cdot} + M \xrightarrow{k_5} Br_2 &
\end{array}$$

M represents all species of molecules in the gas; ΔH_0: standard enthalpy of reaction.

The constant k_1 is very small compared with all the constants which follow, so that a stationary state is quickly established.

The rate of appearance of HBr is

$$\frac{d|HBr|}{dt} = k_2 |Br^{\bullet}| |H_2| + k_3 |H^{\bullet}| |Br_2| - k_4 |H^{\bullet}| |HBr|.$$

Application of the steady-state principle gives for

$$H^0 \rightarrow k_2 |H_2| |Br^{\bullet}| = k_3 |H^{\bullet}| |Br_2| + k_4 |H^{\bullet}| |HBr|$$

for

$$Br^{\bullet} \rightarrow 2k_1 |Br_2| |M| + k_3 |H^{\bullet}| |Br_2|$$
$$+ k_4 |H^{\bullet}| |HBr| = k_2 |Br^{\bullet}| |H_2| + 2k_5 |Br^{\bullet}|^2 |M|$$

Since

$$|Br^{\bullet}| = \left| \frac{k_1}{k_5} \right|^{1/2} |Br_2|^{1/2}$$

with $K = k_1/k_5$ for this dissociation equilibrium, and

$$|H^{\bullet}| = \frac{k_2 |H_2| \left(\dfrac{k_1}{k_5} \right)^{1/2} |Br_2|^{1/2}}{k_3 |Br_2| + k_4 |HBr|}$$

we obtain, by replacing these values in the rate equation

$$\frac{d|HBr|}{dt} = \frac{2k_2 |H_2| \left(\dfrac{k_1}{k_5} \right)^{1/2} |Br_2|^{1/2}}{1 + \dfrac{k_4}{k_3} \dfrac{|HBr|}{|Br_2|}}$$

The constants of Bodenstein's equation are then the following

$$k = \frac{2k_2 k_3 \left(\dfrac{k_1}{k_5}\right)^{1/2}}{k_4} \qquad m = \frac{k_3}{k_4}$$

The enthalpies of activation are

$$\Delta H_1^{\neq} = 46 \text{ kcal mole}^{-1} \qquad \Delta H_4^{\neq} \# 1$$
$$\Delta H_2^{\neq} = 17.5 \qquad\qquad \Delta H_5^{\neq} \# 0$$
$$\Delta H_3^{\neq} \# 1 .$$

C. Synthesis of Hydrochloric Acid

$$H_2 + Cl_2 \rightarrow 2HCl .$$

Many studies have been made of this reaction, both thermal and photochemical, but the experimental results, particularly of the thermal reaction, cannot yet be tied up in as satisfactory a manner as in the preceding cases. In effect, the activation energy of 57 kcal mole^{-1} for the dissociation of molecular chlorine corresponds to a very slow reaction at ordinary temperatures. If the temperature is raised the propagation reactions become very rapid and the great chain length causes an explosion. In the homogeneous gas phase the rate of propagation of the chain is much faster than the rate of homogeneous termination, and a stationary state cannot be established. Furthermore, the walls of the containers have an obvious influence and indicate a heterogeneous terminations process.

Photochemical studies have shown that below 172°C a stationary state is established, but this temperature is too low for the rate of thermal initiation to be sufficient.

However, from the two groups of experiments a mechanism can be advanced

$$Cl_2 \rightarrow 2Cl^\bullet \qquad \text{Initiation}$$

$$\left.\begin{array}{l} Cl^\bullet + H_2 \rightarrow HCl + H^\bullet \\ H^\bullet + Cl_2 \rightarrow HCl + Cl^\bullet \end{array}\right\} \text{Propagation}$$

$$\left.\begin{array}{ll} 2Cl^\bullet + M \rightarrow Cl_2 & \text{Homogeneous} \\ 2Cl^\bullet \xrightarrow{\text{wall}} Cl_2 & \text{Heterogeneous} \end{array}\right\} \text{termination.}$$

These studies are made difficult by the sensitivity of the reaction to traces of foreign products (nitrogeneous substances), which can lead to a fairly long induction period. Oxygen inhibits the reaction by capturing chain carriers. Furthermore, the synthetic reaction is very sensitive to the action of light, particularly blue light, and the thermal reaction must be carried out in the absence of all radiation.

4.5. Reactions in Open Sequences

The different steps of these reactions always proceed in the same order, without repetition of a propagation-step. The steady-state principle is applicable and allows us to give a satisfactory description of them.

EXAMPLE 1. Thermal decomposition of N_2O_5.

$$2N_2O_5 \rightarrow 2N_2O_4 + O_2$$
$$\uparrow \downarrow$$
$$4NO_2$$

This reaction rigorously follows the first-order law, both in the gas phase and in carbon tetrachloride solution.

The rate of decomposition can be deduced from the total variation in pressure (although the equilibrium $N_2O_4^{\bullet} \leftrightarrows 2NO_2$ complicates the calculation) or, when the reaction is carried out in carbon tetrachloride solution, from the partial pressure of oxygen, the other constituents being soluble.

In the first case

$$P_{total} = P_i + pO_2 + 2\alpha pO_2$$

where P_i is the initial pressure and α is the degree of dissociation of N_2O_4. At $t = -\infty$, $P_{final} = (\frac{3}{2} + \alpha) P_i$.

Different mechanisms have been proposed to take account of the experimental first-order kinetics of this reaction, involving a molecule containing more than six atoms, by making use of the estimates of Slater. The most interesting is that of Ogg:

(1) $N_2O_5 \qquad \rightarrow NO_2 + NO_3$
(2) $NO_2 + NO_3 \rightarrow N_2O_5$
(3) $NO_2 + NO_3 \rightarrow NO_2 + O + NO$
(4) $NO + N_2O_5 \rightarrow 3NO_2$.

While the energy of activation E_2 is practically zero, that of step 3 (which is endothermic) is much greater, so that $k_2 \gg k_3$. This step 3 is accordingly rate determining and is bimolecular.

The stationary state applied to NO_3 gives

$$k_1 |N_2O_5| - k_2 |NO_3| |NO_2| - k_3 |NO_3| |NO_2| = 0$$

$$|NO_3| = \frac{k_1 (N_2O_5)}{(k_2 + k_3) NO_2}$$

and the equation for the overall rate becomes

$$-\frac{d |N_2O_5|}{dt} = 2k_3 |NO_3| |NO_2|$$
$$= 2 \frac{k_1 k_3}{k_2 + k_3} |N_2O_5| = k' |N_2O_5|.$$

The ensemble is thus equivalent to a first-order process with an overall energy of activation

$$E = E_1 + E_3 - E_2 = 24.6 \text{ kcal mole}^{-1}$$

with

$$E_1 \# 20, \quad E_2 \# 0 \quad \text{and} \quad E_3 \# 5 \text{ kcal mole}^{-1}.$$

EXAMPLE 2. Ortho-para-hydrogen conversion. Exchange reaction.

At 25°C, hydrogen is composed of 3 parts of *ortho*-hydrogen and one part of *para* (of opposed spins); at low temperature on a catalyst (active charcoal) the gas can be enriched in *para*-hydrogen. On the other hand, at 923K the para is converted into the *ortho* by a reaction of $\frac{3}{2}$ order, which has the following mechanism:

$$\begin{cases} H_2 \to 2H^{\bullet} \\ H^{\bullet} + \underset{para}{H_2} \to H_2 + \underset{ortho}{H^{\bullet}} \quad \text{slow} \\ 2H^{\bullet} + H_2 \to 2H_2 \quad \text{with} \quad p_{H^{\bullet}} = (K p_{H_2})^{1/2} \end{cases}$$

and

$$\mathscr{V} = k p_{H^{\bullet}} \, p_{H_2 para} = k K_{H_2}^{1/2} p^{1/2} p_{H_2 para} \, .$$

With respect to the concentration of atomic hydrogen and to that of *para*-hydrogen, the reaction is second-order, as was shown by Harteck and Wredi.

The reaction $H_2 + D_2 \rightarrow 2HD$ follows a similar mechanism.

4.6. Oxidation Reactions

Oxidations in the gas phase go by a radical chain process, often complicated by branching. They lead to explosions, more or less violent, accompanied by light emission due to 'excited' molecules.

The branching of the chains, by increasing considerably the number of active centres, prevents the establishment of a steady state, because the rate of termination reactions is too low to compensate for the increase in the number of active centres.

Furthermore, the different steps often have a pronounced heterogeneous character, and the state of the surfaces of the containers becomes important. A great number of these oxidation reactions have upper and lower limits separating the explosive and the thermal domains, known as 'explosion forks'.

The general type of oxidation process is the following:

(1) $M \rightarrow R + Products$
(2) $R + M \rightarrow \alpha R + M \, .$
(3) $R + M \rightarrow final \ products$

(4) $R \xrightarrow{\text{wall}}$ Products

(5) $R \xrightarrow{\text{homogeneous}}$ Products

If α is small, a steady state is established, so that

$$\frac{d|R|}{d} = k_1 |M| + k_2 (\alpha - 1) |R| |M|$$
$$- k_3 |R| |M| - (k_4 + k_5) |R| = 0.$$

In this case

$$|R| = \frac{k_1 |M|}{k_3 |M| + k_4 + k_5 - k_2 (\alpha - 1) |M|}$$

and the rate is

$$\mathscr{V} = k_3 |R| |M| = \frac{k_1 k_3 |M|^2}{k_3 |M| + k_4 + k_5 - k_2 (\alpha - 1) |M|}$$

an expression equivalent to

$$\mathscr{V} = \frac{F_i}{F_t + A(1 - \alpha)}$$

where F_i, F_t and A are respectively the rates of initiation, termination, and chain branching.

As long as F_t remains greater than $A(1 - \alpha)$, a steady state prevails, but if $\alpha \gg 1$, the denominator can be positive or negative. Each explosion limit corresponds to a zero denominator threshold of explosiveness.

EXAMPLE 1.   Detonating mixture of $H_2 + O_2$.

The study of a mixture of two parts of hydrogen to one part of oxygen under different conditions of temperature and pressure leads to Figure 13.

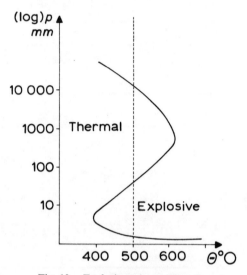

Fig. 13.   Explosive mixture $H_2 + O_2$.

The domain to the left of the curve is thermal, that to the right is explosive.

The lower limit of explosion is heterogeneous, but the second is found in the gas phase and corresponds to an equality between the rate of termination and the rate of branching of the chains.

The mechanisms given to explain these different facts are not all in accord, but indicate for the principal initiation step

$$H_2 \rightarrow 2H^\cdot$$

with a contribution from

$$O_2 \rightarrow 2O^\bullet$$

and

$$H_2 + O_2 \rightarrow HO_2^\bullet + OH^\bullet.$$

The following two steps produce branching:

$$H^\bullet + O_2 \rightarrow OH^\bullet + O^\bullet$$
$$O^\bullet + H_2 \rightarrow OH^\bullet + H^\bullet$$

then the chain is propagated by

$$OH^\bullet + H_2 \rightarrow H_2O + H^\bullet$$
$$HO_2^\bullet + H_2 \rightarrow H_2O + OH^\bullet$$

The termination reactions take place
– in homogeneous gas phase

$$H^\bullet + O_2 + M \rightarrow M + HO_2^\bullet$$

– at the boundary

$$H^\bullet, OH^\bullet, HO_2^\bullet \xrightarrow{\text{wall}} \text{Products}$$

The second limit can be interpreted quantitatively in the following manner: it corresponds to the compensation of chain branching $k_a |H^\bullet| |O_2|$ by trimolecular termination

$$k_t |H^\bullet| |O_2| |H_2| + k'_2 |H^\bullet| |O_2| |O_2|$$

so that

$$k_a = k_t |H_2| + k'_t |O_2|$$

with

$$\frac{k_t}{k_t'} = 0.325.$$

In the presence of an inert gas the homogeneous termination processes are favoured and

$$k_a = k_t |H_2| + k_t' |O_2| + k_t'' |M|.$$

The branching constant k_a varies strongly with temperature, contrariwise to the trimolecular constants k_t, so that the pressures of H_2, O_2 and M corresponding to the second limit vary strongly with temperature.

EXAMPLE 2.   Oxidation of hydrocarbons: case of methane.

The oxidation of hydrocarbons follows very complex ramified processes. In the case of methane, all the mechanisms proposed make use of the radicals $CH_3^•$ and $OH^•$. The most probsble process (Semenov, 1960) involves a branching chain with the formation of formaldehyde, and takes account of latest kinetic data and of modern concepts of the free energy of 'free radical-stable molecule' reactions. Its validity is confirmed by the accord between values calculated for each elementary step and the experimental results.

The initiation step is above 300–350°C

(I) $CH_4 + O_2 \rightarrow CH_3^• + HO_2^• \, (\Delta H = -55 \text{ kcal})$

The propagation and branching steps are

$$(I') \quad CH_3^{\cdot} + O_2 \quad \rightarrow HCHO + OH^{\cdot}$$

$$\begin{cases} (II) & OH^{\cdot} + CH_4 \quad \rightarrow H_2O + CH_3^{\cdot} \\ (II') & OH^{\cdot} + HCHO \rightarrow H_2O + HCO^{\cdot} \end{cases}$$

$$(III) \quad HCHO + O_2 \quad \rightarrow HCO^{\cdot} + HO_2^{\cdot}$$

$$(IV) \quad HCO^{\cdot} + O_2 \quad \rightarrow CO + HO_2^{\cdot}$$

$$\begin{cases} (V) & HO_2^{\cdot} + CH_4 \quad \rightarrow H_2O_2 + CH_3^{\cdot} \\ (V') & HO_2^{\cdot} + HCHO \rightarrow H_2O_2 + HCO^{\cdot} \end{cases}$$

$$\text{Termination } OH^{\cdot} \xrightarrow{\text{wall}} \text{Products}$$

This mechanism is followed by processes for the reaction of carbon monoxide and of hydrogen peroxide, with subsidiary formation of methanol.

The concentration of the active centre $OH^{\cdot}$, a chain propagator, is

$$|OH^{\cdot}| = 2 \, \frac{k_0 \, |CH_4| \, |O_2| + k_3 \, |HCHO| \, |O_2|}{\mathscr{V}_6}.$$

Formaldehyde accumulates according to the equation

$$\frac{d \, |HCHO|}{dt} = k_2 \, |CH_4| \, |OH^{\cdot}|^2 \left[1 - \frac{k_2' k_5' \, |HCHO|^2}{k_2 k_5 \, |CH_4|^2} \right]$$

and reaches its maximum when

$$\frac{d \, |HCHO|}{dt} = 0$$

so that

$$|HCHO|_{max} = \left(\frac{k_2 k_5}{k_2' k_5'} \right)^{1/2} |CH_4|.$$

At the same time, the rate of disappearance of methane is

$$-\frac{d|CH_4|}{dt} = 2\left[k_2|CH_4| + k_2'|HCHO|\right] \frac{k_0|CH_4||O_2| + k_3|HCHO||O_2|}{\mathscr{V}_6}.$$

Towards 500°C, $a_2 \gg a_2'$ and $\overset{\cdot}{a_3} \gg a_0$ (a high rate of branching) so that the maximum rate of decomposition is

$$\left[-\frac{d|CH_4|}{dt}\right]_{max} = \frac{2k_2}{\mathscr{V}_6} k_3|CH_4||O_2||HCHO|_{max}$$

$$= \frac{2k_2k_5}{\mathscr{V}_6}\left[\frac{k_2k_5}{k_2'k_5'}\right]^{1/2}|CH_4|^2|O_2|.$$

This would indicate that the order of the maximum rate with respect to total pressure is three. Experimentally, the total order is 2.7, and with reference to oxygen it is 0.96, in satisfactory agreement with the proposed scheme.

EXAMPLE 3.   Oxidation of carbon monoxide.

The combustion is much influenced by the presence of water vapour and of hydrogen, which facilitate the formation of free radicals in the initiation step. The reaction has been studied by Hinshelwood and Semenov, and is accompanied by one of the most luminous flames known (one quantum of radiation for 125 molecules of CO_2).

The most plausible scheme is the following:

$$CO + O_2 \quad \rightarrow CO_2 + O^{\cdot}$$
$$O^{\cdot} + CO + M \rightarrow CO_2^* + M$$

$$O + O_2 + M \rightarrow O_3 + M$$
$$O_3 + CO \rightarrow CO_2 + 2O^\cdot$$
$$O_3 + CO + M \rightarrow CO_2 + O_2 + M$$
$$CO_2^* + M \rightarrow O^\cdot + CO + M$$
$$CO_2^* + O_2 \rightarrow CO_2 + 2O^\cdot$$
$$CO_2^* \rightarrow CO_2 + h\nu$$
$$CO_2^* + O_2 \rightarrow CO_2 + O_2^*$$
$$O_2^* \rightarrow O_2 + h\nu.$$

The chain carriers are $O^\cdot$ and O_3, and two 'excited' molecules, CO_2^* and O_2^*, are produced and are responsible for the vividness of the flame.

4.7. Molecular Beams

Classical kinetic studies are limited to experiments made on a large number of molecules. In studies of mechanism, it would be very desirable to reach molecular reality, because the knowledge of microscopic properties of reactants enables us to represent the rate constant by a series of terms which are calculable from molecular parameters.

The only experimental attempts in this direction have been carried out by crossed molecular beams, under vacuum conditions sufficiently drastic to avoid parasitic collisions before or after the reaction episode.

A. *Principal*

If the molecular density is very much reduced, the mean free paths of molecules become sufficient for their trajectories to be collimated in rays analogous to light rays (Dunoyer). The beams thus obtained in a high vacuum (10^{-7} to 10^{-8} torr) can be directed with precision.

For this, the molecules coming from a source pass across a series of parallel screens having variable openings (slits), and there are only rare collisions between molecules because of very rapid molecules overtaking the others.

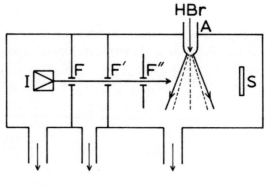

Pumps

Fig. 14.   Crossed molecular beams: F, slits; A, non-collimated beam (HBr); S, detector; I, atom source (K).

To study bimolecular processes, the primary beam is intercepted by another one, which may not be collimated, and which is directed at 90° to the first. Because of the very low pressures the yield is minute, and to obtain measurements sufficiently precise it is necessary to choose reactions of low energy of activation and of high cross section of capture. In any case, the detection of the molecules formed by reactive collisions between the two beams can only be made with extremely sensitive detectors, which has served to limit the development of these studies. In fact, the only detector which has proved satisfactory for this work has been the *surface ionisation detector* of Langmuir, and this responds only to alkali metals and their compounds.

The best-studied reactions have been the following:

(a) interception of molecules of HBr by atoms of potassium, with and without selection of velocities of the molecules.

(b) Interception of molecules of CH_3I by atoms of K, Cs, Rb.

In the first case

$$K + HBr \rightarrow H^\bullet + KBr \quad (E = 3 \text{ kcal mole}^{-1})$$

the atomic beam is obtained by vapourising potassium in a vacuum between 250 and 550°C, while the molecules of hydrobromic acid come from a reservoir between 100 and 200°, by effusion through a very small hole.

The distinction between atoms of K and molecules of KBr is accomplished by replacing the tungsten wire of the detector by a platinum wire which is insensitive to KBr.

The interpretation of the results is complicated by the difference in mass of the two products $H^\bullet$ and KBr, and by the need to assume an average molecular velocity, for a beam containing molecules of very different velocities.

By producing a beam of potassium atoms having all the same velocity (monochromatic), it has been possible to obtain information about molecular parameters, such as the rotational energy of HBr, the value of the potential for the distance of closest approach, the impact parameter, the rotational energy of KBr, etc.

The case of $CH_3I + K \rightarrow CH_3^{\cdot} + KI$ is more favourable because the two products have sizes of the same order. The activation energy is low $(1.3$ kcal mole$^{-1})$, and it has been possible to do experiments at different temperatures and angles of intersection.

The most recent apparatus (STAIR) is capable of producing at least 10^{18} mole cm^{-2}, at energies between 1 and 10 eV. These energies are sufficient, because they correspond to the majority of chemical reactions, and it is much more difficult to accelerate neutral atoms or molecules than charged particles, even to low energies.

The principal chamber is 1.40 m by 4.2 m long; the primary chamber is mounted at right angles, and the primary beam is accelerated by the aid of a supersonic tube, as is the secondary beam. The reactive collisions can be detected by IR spectrometry or by interference spectrometry.

Recent experiments have been most interresting for studies of the reaction process and have already given important results. In particular, it has been impossible to discern any first-order reaction in a single molecular beam, thus furnishing evidence for the necessity of collisions between molecules of the same nature (dissociation of iodine molecules, racemisation of pinene, decomposition of nitrogen pentoxide, etc.).

B. *Ion-Molecule Reactions*

These reactions constitute a particularly interesting field of gas phase kinetics, but although for the particular case

$$H_2^+ + H_2 \rightarrow H_3^+$$

the rate could be calculated theoretically in 1936, it was only about 1950 that the corresponding experiments were started. This delay was due on the one hand to the very rapid progress in the study of radical reactions, which tended to mask the possibility of ionic reactions, and on the other hand to lags in experimental techniques of mass spectrometry. The study of ion-molecule reactions is intimately bound up with the development of this physicochemical technique for detecting chemical species. More particularly, it has been the use of two mass spectrometers in tandem which has led to progress in this field: in the first spectrometer ions of known identity and energy are produced, and are sent into the reaction chamber of the second apparatus, which is used to identify the reaction intermediates and the products formed. In the study of a certain number of cases, it has been found that the reactions have a large capture cross-section.

In a system of mass-spectrometers in tandem, many types of reactions can be observed, such as the exchange mechanism of an electric charge

$$A^+ + B \rightarrow A + B^+$$

which is followed by the dissociation of B, the transfer of a proton, etc.

The spectrometers have been used perpendicularly or in parallel, and recently it has been possible to obtain even the total angular distribution of products.

If the pressure in the collision chamber is low, we obtain the removal of a part of the structure of the ion or molecule. If the pressure is increased, the incident ions ionize the gas molecules by charge exchange, and reactions take place between these ionized gas molecules (or their fragments) and neutral molecules.

EXAMPLE 1.   By bombardment of ethylene by positive ions, we obtain at low pressures the primary ions $C_2H_4^+$; then at higher pressures, secondary ions $C_2H_5^+$, $C_2H_3^+$, $C_2H_2^+$, $C_4H_8^+$, tertiary ions $C_3H_5^+$, $C_4H_7^+$, $C_3H_3^+$, $C_4H_5^+$, $C_4H_6^+$ and quaternary ions $C_5H_9^+$, $C_5H_7^+$.

EXAMPLE 2.   The reaction $O^+ + N_2 \rightarrow NO^+ + N(-25$ kcal mole$^{-1})$ is very important in the study of the ionosphere of the atmosphere; after sunset O^+, produced by photoionization, persists for the whole night. The capture cross-section, measured as a function of the energy of the ion beam, passes through a maximum, equal to $5 A^2 \times (10^{-16} cm^2)$ at 15 eV.

EXAMPLE 3.   In the case of the reaction $O^- + N_2O$, Figure 15 shows the existence of the ion NO^- $(m/e = 30)$ which comes from the reaction $O^- + N_2O \rightarrow NO^- + NO$, this accounts for the effect of pressure, which (for an energy of 2.3 eV) is linear for the peak $m/e = 16$ and is quadratic for $m/e = 30$, with $k \neq 2 \times 10^{-11}$ cm^2 mole^{-1} s^{-1}. Besides these ions, one also finds negative ions of $m/e = 32$, 44, and 46, which are explicable by the secondary reactions

$$
\begin{array}{lll}
O^- + N_2O \rightarrow O_2^- \quad + N_2 & 32 \\
O^- + N_2O \rightarrow N_2O^- + O & 44 \\
O^- + N_2O \rightarrow NO^- \quad + N & 46 \\
\qquad\qquad k \# 4 \times 10^{-13} \text{ cm}^3 \text{ mole}^{-1} \text{ s}^{-1}.
\end{array}
$$

although the origin of the NO_2^- ion is still uncertain, and can also come from pyrolysis of N_2O on the heating filament.

EXAMPLE 4. The study of ion-molecule reactions in flames is important in kinetic theory because a flame presents an environment where the pressure, temperature, and composition can be easily controlled: furthermore, the average energy of an individual molecule (2000 K corresponds to a translational energy of 0.26 eV) in a flame is low in comparison with the cases cited above. The relative simplicity of charged and uncharged species in flames allows us to interpret processes from fundamental parameters and to obtain precise values of rate constants.

Although flame temperatures can vary between 1000 K and 4000 K, most results have been obtained at 2000 K. The proportion of ions in a flame is between 10^{-13} and 10^{-5} (if metallic additives are not introduced), and mass

spectrometry is still the principal method of investigation.
An example of the composition of a flame:

		H_2/air
Ratio	. . . .	0.40
Pressure	. . . .	40 torr
T K	. . . .	1420
N_2	. . . .	0.73
H_2O	. . . .	0.15
H_2	. . . .	10^{-6}
$H^{\cdot}$	. . . .	$< 10^{-6}$
$OH^{\cdot}$	. . . .	8.4×10^{-5}
$O^{\cdot}$	. . . .	1.8×10^{-6}
O_2	. . . .	0.12
NO	. . . .	6.2×10^{-4}

The most simple device consists of a gas-burner fitted
with a grille to assure a uniform distribution of gas; the
distance for observation above the burner gives the time
of the reaction, from the volumetric flow of gas.

EXAMPLE 5. In hydrocarbon flames, many positive and
negative ions have been identified. These are formed by
ion-molecule reactions and the question of the primary
species comes up.

Positive ions → primary reactions

$$CH^{\cdot} + O^{\cdot} \rightarrow CHO^+ + e^- \qquad \Delta H + 20 \text{ kcal mole}^{-1}$$
$$CH^* + C_2H_2 \rightarrow C_3H_3^+ + e^- \qquad \Delta H - 58 \text{ kcal mole}^{-1}.$$

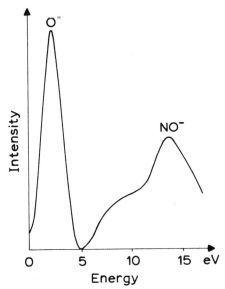

Fig. 15. Reaction $O^- + NO_2 \rightarrow NO^- + NO$.

This ion $C_3H_3^+$ appears in the front of the flame, and reaches a maximum in rich flames.

The negative ions are present in concentrations 100 times smaller and are produced primarily by

$$e^- + H_2O \rightarrow OH^- + H^\cdot$$

whose rate constant has not yet been determined precisely (of the order of 10^{-26} cm^3 mole^{-1} s^{-1} at 2000 K).

The following trimolecular reactions are equally plausible

$$e^- + O_2 \; + M \to O_2^- \; + M \quad k \neq 3 \times 10^{-30}$$
$$e^- + OH^\bullet + M \to OH^- + M \quad k = 10^{-30}$$
$$e^- + O^\bullet \; + M \to O^- \; + M$$
$$e^- + H_2O + H^\bullet \to OH^- + H_2$$

but the first is the least probable because O_2^- appears far from the front of the flame.

All the ions which it has been possible to identify in these combustions (with the exception of CHO^+ and $C_3H_3^+$) are produced by ion-molecule reactions.

The ions CHO^+ are particularly important, and react with the neutral molecules HCOH, CH_3OH, CH_3CHO introduced at a relatively low temperature (600 K) to give species having masses of 31, 33, and 45, according to the reactions

$$CHO^+ + CH_2O \; \to CH_2OH^+ \quad\; + CO$$
$$CHO^+ + CH_3OH \; \to CH_3OH^+ \quad\; + CO$$
$$CHO^+ + CH_3CHO \to CH_3CHOH^+ + CO.$$

CHAPTER 5

REACTIONS IN SOLUTION

5.1. General

The concepts developed in the preceding chapter apply equally to the dynamics of reactions in solution; the formal expressions are similar (justifying the first chapter), because the rates of reaction depend on the frequency of encounter between chemical entities and on relations involving the same energetic or geometric constraints in solution as in the gas phase.

In a 10^{-2} M solution the reactant molecules are separated by a distance equivalent to 20 molecular diameters, which corresponds to the gaseous state under ordinary pressures. But the number of solvent molecules is in large excess (about 1000-fold more), and the reactants being in almost permanent contact with solvent can be charged electrically in the course of various interactions. On the other hand, the ionization of a gas requires considerable energy so that thermal reactions at ordinary or slightly elevated temperatures proceed only by neutral intermediates (homolysis). The formation of ions from the molecules of reactants is much easier in solution because of solvation phenomena, so that the reaction intermediates are often charged (heterolysis).

The comparison of the rate coefficients of a reaction carried out both in the gas phase and in solution is very interesting; Benson could show that the ratio of the coefficients has the value

$$\frac{k_s}{k_g} = \frac{K_s^{\neq}}{K_g^{\neq}}$$

and depends only on the molecularity of the reaction (the number of reactant molecules entering into the active complex), according to the expression

$$\frac{k_s}{k_g} = \frac{n \times 10^{2n-2}}{e^{n-1}},$$

so that

$$n = 1 \ (\text{monomolecularity}) \rightarrow k_s \neq \qquad k_g$$
$$n = 2 \ (\text{bimolecularity}) \quad \rightarrow k_s \neq \quad 100 \, k_g$$
$$n = 3 \ (\text{trimolecularity}) \quad \rightarrow k_s \neq 4000 \, k_g$$

It has been possible to verify these conclusions experimentally for the case where $n = 1$ and in a few instances when $n = 2$. Thus the values of the rate coefficient and of the free energy of activation are identical in the gas phase and in solution for the decomposition of nitrogen pentoxide N_2O_5 and of di-tert-butyl peroxide, and for the dimerization of cyclopentadiene.

The influence of solvent molecules on the dynamic behaviour of solute molecules is very complex. Different effects have been distinguished, of which the following are most important.

5.2. Solvation of Reactants and of the Active Complex

The solvation of reactants and of the active complex by solvent molecules affects the rates of reactions considerably. Solvation energy can be large, and make possible the occurrence of stable charged entities in solvents of high dielectric constant, where they can approach each other closely with corresponding activation energies still low. An example is the electron transfer reaction between 'aquo' complexes of Fe^{2+} and Fe^{3+} (isotopic)

$$Fe(OH_2)_6^{2+} + {}^*Fe(OH_2)_6^{3+}$$
$$\rightarrow Fe(OH_2)_6^{3+} + {}^*Fe(OH_2)_6^{2+}$$

with $E = 9.3$ kcal mole^{-1} and $\Delta G_0^{\neq} = 16.3$ kcal mole^{-1} at $0\,°C$.

These values are compatible with an approach to a distance of $7\,Å$ (the point of contact for the primary solvation shells), for which the coulombic energy of repulsion is only 4 kcal mole^{-1}. (In the gas phase or in a solvent of low dielectric constant, these repulsive interactions would reach 200 kcal mole^{-1} under the same conditions.)

5.3. Solvent = Reactant

If the solvent participates in the reaction, it must be considered a reactant, but it is difficult to demonstrate its role kinetically because, being in large excess, its concentration does not vary appreciably. As a result, the order of the reaction is always degenerate in relation to solvent; in particular, it is difficult to separate its intervention as a reactant from a catalytic action: only chemical analysis of products can distinguish between the two roles.

5.4. Solvent = Donor-Acceptor of Protons

This case is very important in proton-transfer reactions, such as

$$(CH_3)_3NH^+ + (CH_3)_3 \overset{\centerdot}{N} \longrightarrow (CH_3)_3N + (CH_3)_3 \overset{\centerdot}{N}H^+$$

$$\overset{\centerdot}{(CH_3)_3N^+} \cdots H\!-\!\underset{\underset{H}{|}}{O}\!-\!H \cdots \overset{\centerdot}{N}(CH_3)_3$$

5.5. Cage Effect

The duration of an encounter is very different in the liquid phase and in the gas phase, where the probability of two successive collisions between the same pair of molecules is very rare. In solution, after a pair of ions or molecules have diffused together, their separation is kindered by the solvent environment whose molecules constitute a 'cage'. Hence the duration of such an encounter in water is of the order of 10^{-11} s, long enough for several hundred collisions between the two reactant molecules.

For reactions which tend to take place at almost every collision, the rate is limited only by the diffusion process, the coefficient k_D being given by Debye's equation

$$k_D = \frac{4\pi\sigma_{AB}D_{AB}N}{10^3}\left(\frac{w}{e^w - 1}\right)$$

$\sigma_{AB} \rightarrow$ minimum distance between A and B.

D_{AB} mean diffusion coefficient of A and B.

$$w = \frac{Z_A Z_B e^2}{\varepsilon \mathscr{K} T \sigma_{AB}}$$

$\mathscr{K}$ = Boltzmanns' constant,

Z_A, Z_B = Charges on two reactants,

ε = dielectric constant,

e = charge of the electron.

In aqueous solution k_D is of the order of 10^{+10} M^{-1} s^{-1} for molecules and slightly more or less for ions of opposite or the same charge. This value constitutes the limit for rate coefficients of bimolecular processes, which are then diffusion-controlled and have an activation energy of 3 to 5 kcal mole^{-1}. Many proton-transfer reactions are of this type.

EXAMPLES

$$\left.\begin{array}{l} H_3O^+ + OH^- \rightarrow 2H_2O \\ H_9O_4^+ + H_7O_4^- \rightarrow 8H_2O \end{array}\right\} k = 1.4 \times 10^{11} \text{ M}^{-1} \text{ s}^{-1}$$

$$2I^{\bullet} \xrightarrow{CCl_4} I_2 \qquad k = 7 \times 10^9 \text{ M}^{-1} \text{ s}^{-1}.$$

Note: For normal reactions in the gas phase, the collision frequency is from 10^{10} to 10^{11} M^{-1} s^{-1}; because of

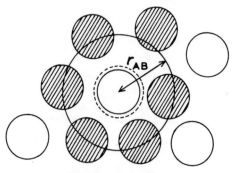

Fig. 16.   Cage effect.

the cage effect, this frequency is increased at least 10-fold in solution, and lies between 10^{11} and 10^{12} M^{-1} s^{-1}.

5.6. Theory of the Activated Complex in Solution

The theory of the activated complex can be proved only for reactants forming a mixture of perfect gases. In particular, the equation

$$\mathscr{V} = \frac{\mathscr{K}T}{h} |X^{\neq}|$$

where the rate is proportional to the concentration of active complex considered as a normal molecule, and where the first term is the universal frequency, cannot be demonstrated directly in the case of real solutions. Experimental verifications can only serve to justify our confidence in applying the theory in liquid media.

In a non-ideal solution, especially when the reactants are ionic species, it is necessary to make use of thermodynamic activities in applying the law of mass action to the activated complex.

$$A^{Z_A} + B^{Z_B} \rightleftarrows (A \ldots B^{\neq})^{Z_A + Z_B} \rightarrow \text{Products}$$

so that

$$K^{\neq} = \frac{(A \ldots B^{\neq})^{Z_A + Z_B}}{(A^{Z_A})(B^{Z_B})}$$

where the parentheses represent activities; the concentration of the active complex becomes

$$|A \ldots B^{\neq}|^{Z_A + Z_B} = K^{\neq} \frac{\gamma_A \gamma_B}{\gamma^{\neq}} |A| \, |B|$$

so that

$$\mathscr{V} = \frac{\mathscr{K}T}{h} K^{\neq} \frac{\gamma_A \gamma_B}{\gamma^{\neq}} |A| \, |B|$$

and the rate coefficient becomes

$$k = \frac{\mathscr{K}T}{h} K^{\neq} \frac{\gamma_A \gamma_B}{\gamma^{\neq}} \, .$$

At infinite dilution the activity coefficients tend to unity and

$$k \rightarrow k_0 = \frac{\mathscr{K}T}{h} K^{\neq}$$

with

$$k = k_0 \frac{\gamma_A \gamma_B}{\gamma^{\neq}} \quad \text{(Bronsted-Bjerrum)}.$$

This relation has been demonstrated in the case of very dilute electrolytes (where the Debye-Hückel theory applies); by replacing the individual activity coefficients by their value as given by the limiting law of Debye, we obtain

$$\log k/k_0 = 2BZ_AZ_B \sqrt{I}$$

with B a constant depending on the temperature and the solvent (0.502 for water at 25°), and I=ionic strength $= \frac{1}{2}\Sigma Z_i^2 c_i$, which implies a linear relationship between the logarithm of the rate coefficient and the square rot of the ionic strength.

All the kinetic consequences of this relation, known under the name of primary kinetic salt effect, have been verified qualitatively and quantitatively. In particular:

If Z_A and Z_B are of the same sign, the slope of the curve is positive.

If Z_A and Z_B are of opposite sign, the slope of the curve is negative.

If one of the reactants is neutral $(Z=0)$, k varies very little with ionic strength.

Even in the absence of added electrolyte when the ionic strength of a solution changes during the course of the reaction, the experimental values of k are no longer constant. This explains why certain authors have described processes in solution having apparently a high order, without taking account of this change in ionic strength. In a certain number of cases the experimental work has been repeated with a known ionic strength kept constant during the reaction, and has shown that the reaction rarely has an order higher than 3.

5.7. Reaction Mechanisms in Solution

A. *Primary Reactions*

For a simple equilibrum reaction in solution for which the rate equation is $A + B \overset{k_f}{\underset{k_r}{\leftrightarrows}} C + D$,

$$-\frac{d|A|}{dt} = k_f |A|\,|B| - k_r |C|\,|D| = +\frac{d|C|}{dt}$$

and

$$K_{eq} = \frac{k_f}{k_r}.$$

The simplest mechanism in solution requires at least the three following steps

$$A + B \overset{k_1}{\underset{k_{-1}}{\rightleftarrows}} (A \dots B) \overset{k_2}{\underset{k_{-2}}{\rightleftarrows}} (C \dots D) \overset{k_3}{\underset{k_{-3}}{\rightleftarrows}} C + D$$

where $(A \dots B)$ and $(C \dots D)$ are collision complexes having a sufficient lifetime, because of the cage effect, to be considered true reaction intermediates. If their concentration is low, the stationary state principle applies and gives

$$k_D = \frac{k_1}{1 + \dfrac{k_1}{k_2} + \dfrac{(k-1)(k-2)}{k_2 k_3}}.$$

FIRST LIMITING CASE

$$\frac{k_{-1}}{k_2} + \frac{k_{-1}k_{-2}}{k_2 k_3} \ll 1 .$$

The forward reaction is diffusion-controlled ($k_f \# k_1$).

SECOND LIMITING CASE

$$\frac{k_{-1}k_{-2}}{k_2 k_3} + 1 \ll \frac{k_{-1}}{k_{-2}} .$$

The transformation of (A... B) into (C... D) is sufficiently slow for the formation of (A... B) to constitute a rapid preequilibrium

$$k_f = \frac{k_1 k_2}{k_{-1}} .$$

This is the most common case.

THIRD LIMITING CASE

$$\frac{k_{-1}}{k_2} + 1 \ll \frac{k_{-1}k_{-2}}{k_2 k_3} .$$

Here the reverse reaction is diffusion-controlled

$$k_r = k_{-3} \qquad k_f = \frac{k_1 k_2 k_3}{k_{-1}k_{-2}} .$$

B. *Real Reactions*

In the case of a reaction involving two reversible steps

$$A \underset{k_{-1}}{\overset{k_1}{\rightleftarrows}} B \underset{k_{-2}}{\overset{k_2}{\rightleftarrows}} C$$

(1) If B accumulates and can be measured directly, that is if

$$k_{-1} \ll k_3 \ll k_1$$

the rate of each step can be obtained separately and the mechanism can be established easily (cf. Figure 17, 1).

(2) If B does not accumulate, it cannot in general be determined directly, but the stationary state principal is applicable to it, and allows us to establish the important rate equations for the different elementary steps.

In this case $k_1 \ll k_{-1} + k_2$ (cf. Figure 17, 2).

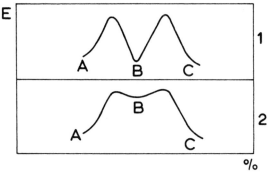

Fig. 17.   Reaction A $\rightleftarrows$ B $\rightleftarrows$ C.

(3) Equally frequently, the reactants form compounds which may or may not be reaction intermediates

(1)   A + B → Products
(2)   A + B $\rightleftarrows$ X Compound
(3)   A + B $\rightleftarrows$ X → P.

If reactions 1 and 2 are competitive and reaction 3 successive, the two mechanisms 1 + 2 and 3 can be distinguished kinetically only if the rate process 1 is very slow beside that of 2. If not, the only way to determine if a rapidly formed compound (or complex) is a reaction intermediate is to follow the reaction in the transitory state immediately after mixing by the methods of fast kinetics. In effect, in the two cases above the transition states have, except for molecules of solvent, the same composition if not the same structure.

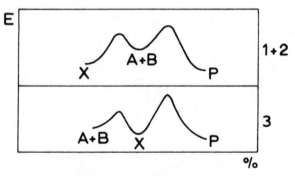

Fig. 18.   Intermediate and complex.

The comparison of the entropy of formation of the supposed intermediate complex with the entropy of activation for the formation of the active complex deduced from kinetic measurements, can give useful indications on the similarity of the structures.

5.8. Reaction Intermediates and Complexes in Solution

The detection of very reactive intermediates can be done in general with the aid of 'scavengers', which react with them to form products, whose isolation constitutes a proof of their existence, or which result in an induced reaction of different stoichiometry (action of I^- on $OH^•$).

In the case of solutions, intermediates can form in the interior of a cage and react before they have time to diffuse out; it is not possible then to identify them with 'scavengers'. The cage effect is also responsible for the poor yield in the photochemical dissociation of halogen molecules in solution, because of recombination in the cage of the pair of free atoms formed.

The methods of fast kinetics (Chapter 3) allow us to detect them, however, because the lifetime of these intermediates in solution, although very small (as shown in the following calculation) is enough to make them accessible to modern techniques. For the most rapid reactions in solution of the type $2A \rightarrow$ Products the time of half-reaction is

$$t_{1/2} = \frac{1}{ka} \quad \text{so that with} \quad k \neq 10^{+10}$$

even if the initial concentration is as low as 10^{-6} M, $t_{1/2}$ is still 10^{-4} s.

The cage effect also makes it difficult to demonstrate the existence of unstable complexes, such as those which could exist between NH_3 and H_2O. In effect, two molecules, H_2O and NH_3, can be found in the same cage for a time, relatively long on the molecular scale, of 10^{-12} s, without forming a hydrogen-bonded complex. Only an experimental demonstration of a longer lifetime would give proof of the existence of a complex. Recent experiments show that this is not the case, unlike that of trimethylamine $(CH_3)_3N$, which forms a complex with water having a lifetime of 10^{-10} s.

5.9. Reactions Between Neutral Molecules

The effect of solvent is limited then to a cage effect and to solvation: the first increases the rate coefficients of bimolecular reactions, and becomes even more important as the size of solvent molecules becomes greater than that of the reactants; it is also increased by increasing viscosity of the solvent.

Weakly-solvating solvents have only a small effect on the rates of reactions (ex. case of N_2O_5 in gas phase and in solution in CCl_4). On the other hand, in strongly-solvating solvents the active complex is often more solvated than the reactants, and this increases the rate of reaction (dimethyl sulfoxide). In general, neutral molecules in solution are not much affected by these interactions. Van der Waals forces are exerted at only very short distances, because they diminish with $1/d^7$, so that the interaction of two molecules cannot exceed $\mathcal{K}T$ (0.6 kcal mole^{-1} at 25°C).

5.10. Ionic Reactions

The electrostatic energy diminishes with the first power of the distance, so that a charged particle can influence an ion separated by several molecular diameters.

For a separation of 10 Å, the interaction is still equal to 100 $\mathscr{K}T$, and even that between an ion and a dipole (energy $= f(1/d^2)$) is still of the order of 20 $\mathscr{K}T$).

These values explain the impossibility of observing ionic equilibria in the gas phase at ordinary temperatures; the existence of individual ions in solution is possible only because they are stabilised by solvation.

The free energy of activation of an ionic reaction results from the three following terms

$$\Delta G^{\neq} = \Delta G_{\infty}^{\neq} + \Delta G_{c}^{\neq} + \Delta G_{\mu}^{\neq} .$$

(a) $\Delta G_{\infty}^{\neq}$ represents the free energy of activation in the absence of any electrostatic effect. One can consider that this term corresponds to a medium of infinite dialectric constant in which $\Delta G_{c}^{\neq}$ and $\Delta G_{\mu}^{\neq} \to 0$.

(b) $\Delta G_{c}^{\neq}$ is the change in free energy corresponding to the coulombic forces brought into play by the collision of two charged reactants.

If the ionic strength of the solution remains low,

$$\Delta G_{c}^{\neq} = W = -\int_{\infty}^{r^{\neq}} \frac{Z_A Z_B e^2}{\varepsilon x^2} \, dx ,$$

which gives

$$\text{Log} \, k = \underset{\varepsilon \to \infty}{\text{Log} \, k_0} - \frac{Z_A Z_B e^2}{\mathscr{K} T \varepsilon r^{\neq}} .$$

The variation of the rate coefficient with the reciprocal of the dielectric constant of the solvent, ε, is often followed very exactly, whether the ions A and B have charges Z_A, Z_B of the same sign or opposite sign.

Remarks: (1) However, the change of solvent, by its very nature, can alter $\Delta G_\infty^{\neq}$ and the contact distance $r^{\neq}$, and so perturb the appplication of the formula above.

(2) The preceding model implies a medium of continuous structure, so that in mixtures of solvents the equation applies more poorly because of the phenomenon of preferential solvation by one of the two solvents. The ionic reaction proceeds more as if it were being carried out in the solvent having the higher dialectric constant.

(3) When the dialectric constant of a liquid is less than $\varepsilon = 20$, it is no longer possible to ignore association to form ion pairs, either in contact (intimate) or bridged by a molecule of solvent, having a reactivity sometimes much diminished.

(c) $\Delta G_\mu^{\neq}$ corresponds to the electrostatic forces brought into play by encounters between reactants and all ionic species of the solution (the ionic atmosphere). This effect, named 'primary kinetic salt effect' has been treated in Section 6.

The rate coefficient of an ionic reaction is then expressed by the following general formula

$$\mathrm{Log}\, k = \mathrm{Log}\, k_0 - \frac{Z_A Z_B e^2}{\mathscr{K} T \varepsilon r^{\neq}} + \frac{Z_A Z_B e^{5/2}}{(\mathscr{K} T \varepsilon)^{3/2}} \left(\frac{8\pi N}{1000}\right)^{1/2} I^{1/2}$$

the Bronsted-Christiansen-Scatchard equation, for which

the Bronsted equation $\log k = \log k_0' + 2BZ_A Z_B I^{1/2}$ (k_0' at zero ionic strength in a solvent of dielectric constant ε) is only a particular case.

Because of the electrostatic model employed (Debye-Hückel) these formulae apply only in solutions having an ionic strength less than 0.02 M and for non-associated ions.

Any two ions having an interaction energy greater than $2\mathscr{K}T$ cannot be considered as free ions but as an ion pair; this energy corresponds to a critical distance of 20 Å in water.

5.11. Reaction Between Polar Molecules (Dipole-Dipole)

Taking account of the free energy of solvation of a sphere of radius r having a dipole moment μ_i

$$\Delta G_i = - \frac{\mu_i^2}{r^3} \frac{\varepsilon - 1}{2\varepsilon + 1}$$

and neglecting the effects of the ionic atmosphere, we can express the effect of solvent on this type of reaction by

$$\text{Log}\, k = \text{Log}\, k_0 + \frac{1}{\mathscr{K}T} \frac{\varepsilon - 1}{2\varepsilon + 1} \left(\frac{\mu_2^{\neq}}{r_3^{\neq}} - \frac{\mu_A^2}{r_A^3} - \frac{\mu_B^2}{r_B^3} \right)$$

where k_0 is the rate coefficient in a reference solvent having $\varepsilon = 1$ and where $\mu^{\neq}$, μ_A and μ_B are the dipole moments of the active complex and of reactants A and B.

According to this formula, due to Kirkwood, the rate coefficient increases with dielectric constant if the active complex is more polar than the reactants. But the effect of solvent on these reactions remains weak, and the formula has limited applicability, because van der Waals forces are no longer negligible beside these weak interactions.

5.12. Secondary Kinetic Salt Effects

In the case of ionic equilibria, the concentration of an ionic species can depend on the ionic strength of the solution. Thus in the case of a weak monoacid ionizing thus

$$AH \overset{S}{\rightleftarrows} H_S^+ + A_S^-$$

the logarithm of the concentration of solvated protons is given by

$$\log |H^+| = \log K_a + \log \frac{(AH)}{(A_S^-)} + 2BI^{1/2}.$$

This effect can be important and sometimes stronger than the primary effect; if it operates in an opposite direction, it can compensate for it to such an extent that the ionic reaction is no longer affected by the ionic strength of the medium.

5.13. Concentrated Electrolyte Solutions

When the concentration of an ionic reactant is greater than 5×10^{-2} M, it becomes necessary to take account of its thermodynamic activity rather than its concentration.

Thus, for nucleophilic substitutions by methoxide ions in methanol solution, Schaal and Terrier showed that the rates of the reaction varied in the ratio of 1 to 10^6 when the concentration of potassium methoxide went from 10^{-2} to 5.6 M, the limiting solubility of the reactant. Furthermore, sodium and lithium methoxide solutions having the same concentration as potassium methoxide solutions are much less reactive.

The rate of reaction remains proportional to the activity

TABLE III

|CH_3OM|	$a_{CH_3O^-}$		
	K	Na	Li
0.2	0.270	0.260	0.186
0.4	0.650	0.589	0.371
0.6	1.27	1	0.590
1	4.17	2.63	1.20
1.4	13.49	6.31	2.10
2.6	389	89.1	
3.	1175	240	
3.6	6457	1096	
4	2×10^4	2.5×10^3	
4.5	8.3×10^4	8×10^3	
5.1	4.46×10^5	2.4×10^4	
5.5	1.35×10^6	4×10^4	

of methoxide ion (CH_3O^-) in the three solutions, as can be deduced from the acidity function H_M of the methanol solutions which is indicated in Table III.

EXAMPLE:

Dinitronaphthalene reacts with potassium, sodium and lithium methoxides to give 1-methoxy-4-nitronaphthalene at different rates when the concentration b of the three reactants in methanol is the same. The interpretation of the mechanism would be difficult, were it not for the fact that use of the activity of the nucleophilic reactant (CH_3O^-) given in the table above leads to a simple linear relationship

$$\mathscr{V} = k_0 \, |\text{dinitronaphtalene}| \, a_{CH_3O^-}$$

This is shown in Table IV, where the pseudo-first-order

TABLE IV

b_{CH_3OK}	a_{CH_3OK}	k_{min}^{-1}
0.14	0.177	0.014
0.29	0.44	0.0325
0.46	0.8	0.072
0.6	1.23	0.102
0.78	2.1	0.207
0.95	3.24	0.35
1.07	4.9	0.46
1.21	7.4	0.7
1.33	10.2	1.07
1.36	11.2	1.035
1.50	16.5	1.5
1.66	26	2.85

$k_0 = 9 \times 10^{-2}$ min^{-1} l^{+1} mole^{-1}

rate coefficient k $(k_0 a_{CH_3O^-})$ is shown as a function of the activity $a_{CH_3O^-}$ of methoxide ion.

5.14. Some Special Mechanisms

The mechanisms of certain families of reactions in solution are particularly important and will be explained briefly in this sestion.

A. *Electron-Transfer Reactions REDOX*

These relatively rapid reactions take place with a transition state including both reductant and oxidant. The simplest, 'electron-exchange reactions', take place between two ions of the same nature but having different oxidation states, and can be followed with the aid of isotopic tracers. For example

$$*Fe^{3+} + Fe^{2+} \rightarrow *Fe^{2+} + Fe^{3+}.$$

In general, two distinct types of processes have been recognized, due to the fact that the ions in solution are almost always surrounded by two layers of solvent molecules. In particular, in the case of a transition element such as cobalt, the first hydration sphere or 'inner shell' corresponds to ligands fixed by coordination bonds and the second, outer sphere includes solvation molecules fixed less rigidly by electrostatic attraction. If, in the active complex, only the outer spheres overlap, the process takes place by the 'outer sphere', but if the inner spheres overlap, it takes place by the 'inner coordination sphere'.

(1) *Mechanism by inner coordination sphere.* The meeting of oxidant and reductant in a transition state causes a modification of the coordination spheres and the two reactant centres become bound by a 'bridge' formed by one of the ligands, without implying the transfer of this ligand from the oxidant to the reductant. This mechanism allows for one or several steps, according to whether the preceding complex is an activated complex or a true reaction intermediate, but because of the very low activation energy for the transformation of the intermediate, the elucidation of the detailed reaction path is difficult.

EXAMPLE. Reduction of a complex of cobalt^{III} by an ion Cr^{2+} in aqueous medium.

The reduction reaction takes place by the following mechanism, involving the intermediacy of a bridge between the two centres formed by the chlorine ligand

$$[Co(NH_3)_5Cl]^{2+} + Cr^{2+}(H_2O) + 5H^+$$
$$\rightarrow (NH_3)_5Co \ldots Cl \ldots Cr(H_2O_5)]^{4+}$$
$$\rightarrow Co^{2+}(H_2O) + [CrCl(H_2O)_5]^{2+} + 5NH_4^+ .$$

The 'bridge' reduces the electrostatic repulsion forces between the two ions of the same charge and allows the delocalization of the metallic orbitals.

(2) *Mechanism by outer sphere.* The coordination sphere in these cases remains intact and only the outer solvation spheres overlap. In this case the rate law corresponds to

an active complex which contains the ions with all their ligands.

The most rapid oxidation-reduction reactions probably take place by this process which involves neither the rupture nor the formation of bonds but, in spite of the great mobility of electrons, the speed of these reactions remains nevertheless limited. In the same way, the mechanism of redox reactions bringing into play several electrons is considered possibly to involve a double bridge.

This process is much less sensitive to the nature of the ligand than the first, and almost certainly requires solvent molecules to act as ligand-bridges, although it is difficult to establish this point.

Studies of the mechanism of electron-transfer have developed particularly with the advent of new methods for fast kinetics.

B. *Substitution Reactions*

In both organic and inorganic chemistry there are many nucleophilic substitution reactions of the general type

$$RX + Y \rightarrow RY + X$$

in which the displaced group X leaves the molecule and the group Y enters.

These reactions differ from electron-transfers by the fact that the entering group forms a bond with the substrate of the same nature as the leaving group.

While certain reactions take place in solution by a homolytic process involving free radicals as in the gas phase

$$R:X + Y^{\bullet} \rightarrow R:Y + X^{\bullet}$$

the greatest number takes place by a heterolytic process involving ions, being either *nucleophilic* substitutions

$$R:X + :Y \rightarrow R:Y + :X$$

or *electrophilic* substitutions

$$R:X + Y \rightarrow R:Y + X.$$

In the first case, $:Y$ is a nucleophilic reactant, donating electrons, which attack the molecules RX at a point of low electron density (reactions denoted SN).

In the second case, the electrophilic reactant Y, having a low-energy orbital vacant, attacks the molecule RX at a point of high electron density.

(1) *Nucleophilic substitution.* These substitution reactions take place by a dissociation (type SN_1) or association mechanism (type SN_2).

(a) *Type* SN_1 *monomolecular* $RX \rightarrow R^+ + X^-$.

The rupture of a bond with the appearance of a charged intermediate (which can be demonstrated directly) precedes the formation of another bond, and can take place only through the intermediacy of molecules of solvent, because the solvation energy compensates in part for the energy necessary to stretch the RX bond.

EXAMPLE: Tertiary alkyl halides.

$$(CH_3)_3CCl + OH^- \rightarrow (CH_3)_3COH + Cl^- \quad \text{overall}$$
$$(CH_3)_3CCl \qquad \rightarrow (CH_3)_3C^+ + Cl^- \quad \text{slow}$$
$$OH^- + (CH_3)_3C^+ \rightarrow (CH_3)_3COH \qquad \text{fast}$$
$$\left. \right\} \text{two steps}$$

The rate of the overall reaction is first order with respect to RX, while the concentration of the reactant nucleophile OH^- has no effect.

$\mathscr{V} = k_1(CH_3)_3CCl$ because the first step (monomolecular) is rate-determining.

These reactions are not stereospecific and generally lead to a racemic mixture of isomers (when the reactant RX is optically active). The intermediate ion is planar; and the reactant can attack as easily from one side as from the other:

(b) *Type* SN_2 *(bimolecular)*.

In this association process, the first bond breaks while the second is being formed in a concerted one-step mechanism. The structure of the transition state is $(Y \ldots R \ldots X)$ with the coordination number greater than that of the molecule RX.

EXAMPLE 1: Substitution of the square-planar complexes of platinum-II

$$PtL_3X + Y \rightarrow PtL_3Y + X.$$

EXAMPLE 2: Substitution of primary alkyl halides

$$CH_3Cl + NH_3 \rightarrow H_3N^+ - CH_3 + Cl^-$$

methylammonium ion.

Two experimental arguments support the SN_2 mechanism:

(a) *Kinetic*: the rate law is first-order with respect to both reactants

$$\mathscr{V} = k_2 |NH_3| \, |CH_3Cl|$$

(b) *Stereochemical*: these reactions are stereospecific and take place either with retention of configuration or with Walden inversion.

For the square-planar complexes of platinum, the transition state has the form of a trigonal bipyramid, with the two groups X, Y in equatorial positions:

A trans complex gives only a trans product, and in the same way a cis only a cis.

In the case of the alkyl halides, when the reactant RX is optically active, bimolecular substitution leads to an optically active product but having inverted configuration. This inversion is explained by the configuration of the transition state.

The three bonds C—A, C—B, C—D, repulsed by the nucleophile Y, become planar in the transition state and then return to a tetrahedral configuration little by little as L moves away.

(2) *Electrophilic substitutions.* Unsaturated compounds (olefins and aromatics) are very sensitive to electrophilic reactants in the presence of acid catalysts. Benzenoid hydrocarbons undergo substitution of the following type:

EXAMPLE: Nitration of benzene by a mixture of nitric and sulfuric acids.

$$HSO_4^- + H_2O + NO_2^+ + C_6H_6 \rightarrow C_6H_5NO_2 + H^+$$

nitronium
ion

$$H^+ + HSO_4^- \rightarrow H_2SO_4.$$

(3) *Reaction intermediates in solution.* Heterolytic processes are common, and the intermediates are most often charged.

Anions: Ex: Organic carbanions of pyramidal structure, resembling that of amines.

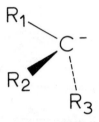

Cations: Carbonium ions of planar structure, in which the central carbon atom is sp^2-hybridized.

$$\begin{array}{c} CH_3 \\ | \\ C^+ \\ CH_3 \diagdown CH_3 \end{array}$$

Complex ions: Ex.: $[(NH_3)_5Co—Cl—Cr(H_2O_5)]^{2+}$, bridged by the chlorine ligand.

But even these heterolytic processes can involve neutral molecules, where one of the atoms is not in its usual valence state.

EXAMPLE: Dichlorocarbene, $:CCl_2$, is formed in the decomposition of chloroform in an alkaline medium.

$$CHCl_3 + OH^- \rightarrow CCl_3^- + H_2O \quad \text{(fast)}$$
$$CCl_3^- \rightarrow CCl_2 + Cl^- \quad \text{(slow)}$$
$$\downarrow$$
$$\text{Products}.$$

C. *Proton-Transfer Reactions*

These reactions are very rapid because the donor does not

need to come in contact with the acceptor; the interposed molecules of solvent can serve to transport the atom of hydrogen by a rearrangement of hydrogen bonds

$$A-H\ldots \underset{\underset{H}{|}}{O}-H\ldots \underset{\underset{H}{|}}{O}-H\ \ldots B$$

$$\rightleftarrows A^- \ldots H-\underset{\underset{H}{|}}{O}\ldots H-\underset{\underset{H}{|}}{O}\ldots H-B^+ .$$

(Grotthus)

The proton does not possess peripheral electrons and is about 10^5 times smaller than the other cations; there is then no preferential orientation in protonic reactions. Its very small size makes it a very strong polarizing agent, so that it is always strongly solvated in solution. In water it exists in the form of the hydronium ion H_3O^+ by formation of a covalent bond to oxygen, followed by solvation by three molecules of water to give the relatively stable $H_9O_4^+$.

Although the rates of protonic reactions are very great, they can be determined by the modern relaxation techniques of Eigen, which allow us to distinguish three steps

(1) $HX + Y \rightleftarrows (XH, Y)$

in which acid and base reactants diffuse towards each other so that a hydrogen bridge can form

(2) $(XH, Y) \rightleftarrows (YH, X)$

in which the proton passes from X to Y

(3) $(YH, X) \rightleftarrows X + HY$

in which the products separate by diffusion.

The overall rate of this process corresponds to the limiting rate of diffusion 10^{11} l mole^{-1} s^{-1}, when the formation of the hydrogen bridge is easy. On the other hand, for acids of the type CH (pseudo-acids), the hydrogen bridge (CH... O) probably does not exist and the overall reaction is 10^5 slower than for the groups OH... O, OH... N, NH... O, NH... N, etc.

Proton-transfer reactions are very important in the study of acid-base catalysis (Chapter 6).

5.15. Thermodynamic and Kinetic Correlations

One of the essential aims of chemical kinetics is to relate the rate or the extent of a reaction to the structure of the reactants; chemical thermodynamics is intimately bound up with these attempts, and even if the calculation of rate and equilibrium constants from structural considerations is not yet possible, it has proved interesting to correlate the reaction parameters with properties of the reactants by empirical relations.

In a homologous series thermodynamic correlations are often discovered; in kinetics, similar correlations are found for families of reactions.

A. *Polanyi's Relationship*

It is found that changes in the activation energy and the heat of reaction are related by an empirical equation.

$$\Delta E = \alpha \Delta (\Delta H)$$

where $0 < \alpha < 1$ depends on the family.

This relationship, equivalent to

$$E = E_0 + \alpha \Delta H$$

applies to exothermic steps and in particular allows us to predict the activation energies of 'molecule + free radical' reactions, when the heats of reaction are known, because Semenov has shown that

$$E_0 = 11.5 \text{ kcal mole}^{-1} \quad \text{and} \quad \alpha = 0.25$$

for a large number of these reactions. These heats of reactions can be deduced, for free radicals, from tables of bond energies.

Polanyi's relationship has been verified for a certain number of cases; its only limitation seems to be with reactions where steric kindrance contributes strongly to the activation barrier.

B. *Linear Free Energy Relationships*

The equation of Polanyi can be put in the following form

$$E = \text{const} + \alpha \Delta H$$
$$E/RT = \text{const} + \frac{\alpha \Delta H}{RT}$$
$$e^{(-E/RT)} = \text{const } e^{(-\alpha \Delta H/RT)}$$

or

$$k = \text{const } K^{\alpha}$$

but Hammett used the letter ρ in place of α and this notation has remained in usage. He showed that for a family of reactions (esterification of ethanol by variously-substituted benzoic acids) the rates of esterification are related to the equilibrium constants K by the relation

$$\log k/k_0 = \rho_k (\log K/K_0)$$

where the index 0 designates the unsubstituted benzoic acid.

Taking another family of reactions for the same benzoic acids, and assuming that a linear relation exists between standard free energies so that

$$^1\Delta G - {}^1\Delta G_0 = \rho_K ({}^2\Delta G - {}^2\Delta G_0)$$

(which in fact cannot be demonstrated), Hammett arrived at a number of empirical correlations which have had much use, but which apply only to *meta* and *para* derivatives.

In effect, in going from ΔG to equilibrium constants we obtain

$$\log {}^1K - \log {}^1K_0 = \rho_K (\log {}^2K - \log {}^2K_0).$$

The difference $\sigma = \log {}^2K_0 - \log {}^2K_0$, for a reaction family taken as a standard (for example: the ionization in water of the substituted benzoic acids) is a constant which depends only on the nature and position of the substituant, so that

$$\log \frac{{}^1K}{{}^1K_0} = \rho_K \sigma$$

which leads to the Hammett relation by using Polanyi's

relation

$$\boxed{\log k/k_0 - \mu\sigma} \quad \text{with} \quad \sigma = \rho_k\rho_K.$$

The constant ρ represents the electronic influence of meta and para substituents on the family of reactions and depends only on the nature of the reaction.

In practice, values of σ are not rigorously constant for all reactions.

Other equations based on linear free energy relations have been proposed, such as that of Taft for ortho derivatives, four variables being used to separate inductive and mesomeric effects

$$\log k/k_0 = \sigma_1\rho_1 + \sigma_2\rho_2,$$

or that of Swain and Scott for showing the nucleophilicity and electrophilicity of reactants

$$\log k/k_0 = ns + es'.$$

C. Grunwald-Winstein Relationship

Grunwald and Winstein tried to relate the rates of solvolysis reactions to the properties of the solvent when the process is of the SN_1 type involving ionization of the reactant. The ionizing power of a solvent is measured by the quantity

$$Y = (\log k/k_0)_{t-BuCl}$$

where k corresponds to the rate coefficient for the solvolysis of tert-butyl chloride in the solvent being considered and k_0 that in the reference mixture H_2O-ethanol (80%). For other alkyl halides the empirical relationship

$$\boxed{\log k/k_0 = mY}$$

has been verified, m being a parameter depending on the substrate (sensitivity of solvolysis of substrate to change in ionizing power of the solvent).

This relationship applies also to some SN_2 reactions. The Y parameter seems even to have a real physical significance, because it correlates remarkably well with the frequency of charge-transfer absorption bands in a large number of solvents. It is however often necessary to take account of association into ion pairs in applying this relationship.

D. *Compensation Effect*

The Hammett correlations imply that differences in rates are due to variations in $\Delta H_0^{\neq}$ (or E_a), while the frequency factor A remains constant for members of the same family. However, the contrary case can be demonstrated (E = = const, A variable), for example, in a reaction carried out in a series of solvents.

In solution, it is possible that E and A are both variable and sometimes even in the opposite directions, so that the effect on the rate coefficient remains slight by compensation.

The most important consequence of this effect is then the appearance of an 'isokinetic' temperature at which all reactions of the same family have the same rate (Figure 19), so that it is essential to measure rate coefficients over a temperature range as great as possible before making any comparison.

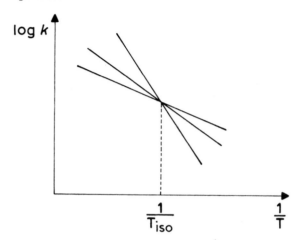

Fig. 19.   Compensation effect.

CHAPTER 6

HOMOGENEOUS CATALYTIC REACTIONS

6.1. General

The use of the term 'catalytic process' was proposed by Berzelius in 1836 to describe the increase in the rate of certain chemical reactions under the influence of substances apparently foreign to the process, such as acids in the hydrolysis of starch. The long-known phenomenon has given rise to many researches and is as important in homogeneous phases as in heterogeneous or enzyme kinetics.

A. *General Considerations*

Any substance which increases the rate of a chemical reaction and which is found with the final products without appearing in the overall stoichiometry is a *catalyst* for the reaction. The mode of action of these catalysts has given rise to much discussion and it seems that there exist two different types.

In many cases, particularly in a homogeneous phase, the catalyst participates in one step of the process, giving a complex with the reactants, and is regenerated in a later step. The reaction mechanism in the presence of a catalyst is clearly different from the simple homogeneous process, and has a lower activation energy. This mode of action is especially frequent when the catalyst molecule is small, thus explaining the importance of acid and base catalysis by the ions H^+ and OH^-.

But the change in the rate of the reaction can be due simply to a change in the environment of the reactants (change in dielectric constant, kindrance to free rotation of atoms, etc.), without which the homogeneous process does not take place.

Remark: The present tendency in catalysis is to extend the notion of catalyst to all substances capable of modifying the rate of a chemical reaction, whether or not they appear in the products.

In effect, a catalyst can catalyze the slow step of a process, be regenerated, and then be consumed in subsequent fast steps, thus entering into the overall stoichiometry.

EXAMPLE. The two following hydrolysis reactions

$$\text{ester} \rightarrow CH_3COOCH_3 + H_2O + H^+$$
$$\rightarrow CH_3COOH + CH_3OH + H^+$$
$$\text{amide} \rightarrow CH_3CONH_2 + H_2O + H^+$$
$$\rightarrow CH_3COOH + NH_4^+$$

are both catalyzed by H^+ ions and are very similar; only the first represents pure catalysis.

The catalytic reaction presents a certain number of characteristics which enable us to classify the role of the catalyst.

(a) A small quantity of catalyst is sufficient to transform a large quantity of reactants, since it turns up unchanged (in pure catalysis) in the final products.

(b) A catalyst does not bring energy to the system reactants-products, and cannot bring about a reaction which corresponds to an increase in free energy: it is necessary that the reaction be thermodynamically possible, even if it does not proceed spontaneously.

The distinction between 'initiation' or 'acceleration' of an infinitely slow reaction is purely academic, and Ostwald's conception seems preferable: the catalyst accelerates a reaction whose rate, in its absence, would be too slow to be measured. It is often possible, on the other hand, to measure the rate corresponding to the homogeneous reaction by itself, and then its rate in the presence of catalyst.

(c) The rate of homogeneous catalytic reactions is proportional to the concentration of catalyst, except in the case of chain processes (Section 9) and ionic reactions in solutions of high ionic strength, where it is necessary to use the thermodynamic activity of the catalyst and not its stoichiometric concentration.

For a reaction $A + B \rightarrow$ Products, capable of being catalyzed by C, the rate equation will be

$$\mathscr{V} = k_0 |A| |B| + k_1 |C| |A| |B| = [k_0 + k_1 |C|] |A| |B|.$$

The rate coefficient of the real process in the presence of catalyst will then be

$$k = k_0 + k_1 |C|$$

where k_0 is the constant of the purely homogeneous process and k_1, the catalytic coefficient of the substance C, is characteristic of the influence of the catalyst on the reaction.

(d) If the catalyst does not participate in the overall stoichiometry of the reaction it cannot, according to the second principle, modify the thermodynamic equilibrium. It follows that a catalyst will influence quantitatively both the forward and the reverse reactions to the same extent, as has been shown experimentally. In the case where the catalyst participated in the stoichiometric balance it can modify the equilibrium only by altering its thermodynamic activity in the course of the reaction.

Figure 20 represents the variation ΔE of energies of activation of the forward and reverse reactions in the

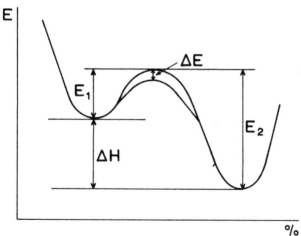

Fig. 20.   Catalysis and activation energy.

presence of catalyst, which does not affect the enthalpy of the reaction.

Catalytic reactions are frequent in the gas phase but above all in solution, where it is often difficult to state precisely the role played by a molecule of solvent or of a constituent of a mixture of liquids. Even in the case when chemical analysis shows that the solvent does not participate in the reaction as a reactant, it is always difficult to decide whether the solvent influences the reaction rate by its physical properties (Chapter 4) or whether it really catalyzes the reaction.

In the case of a change of reaction mechanism in the presence of catalyst, an intermediate complex is formed between catalyst and at least one of the reactants, and this reactant in turn enters into reaction with the other reactant or decomposes spontaneously.

In the case of chain processes, the catalyst acts by accelerating the initiation reaction, that is to say, in favouring the appearance of free radicals to carry the chain.

The following general scheme applies to most catalytic processes outside of closed sequences.

The mechanism of the reaction

$$S + W \rightarrow P + Z$$

in the presence of catalyst C becomes:

$$S + C \underset{k_{-1}}{\overset{k_1}{\rightleftharpoons}} X + Y$$
$$X + W \overset{k_2}{\rightarrow} P + Z + C$$

neglecting the constant k_{-2} of the reverse reaction (-2). If Y and W do not exist the reactions (-1) and (2) then follow the first order law. On the other hand, Y and W are particularly important for acid-base catalysis.

The kinetic expressions depend in fact on the stability of the intermediate X:

(1) If the rate of decomposition (2) is lower than the rate of return (-1), the concentration of complex can be calculated by taking account of the rapid preequilibrium and neglecting reaction (2). X then corresponds to Arrhénius' intermediate.

(2) If on the contrary the rate of formation (1) of the complex is low (rate-determining step), its concentration remains always small; the stationary state principle is then applicable to this intermediate, which corresponds to that of van't Hoff.

Rapid preequilibrium: neglecting the rate of the slow reaction (2), the mass action law gives

$$k_1/k_{-1} = K = \frac{|X|\,|Y|}{|C|\,|S|}.$$

The actual concentration $|C|$ is equal to $|C_0|-|X|$ and $|S| = |S_0|-|X|$ (C_0 and S_0 being initial concentrations), whence

$$K = \frac{|X|\,|Y|}{[|C_0|-|X|]\,[|S_0|-|X|]}$$

which allows us to calculate $|X|$ and to use this value in the expression for the overall rate of the process, which is that of the slow step (2) $\mathscr{V} = k_2 |X| |W|$. But operating conditions often allow this calculation to be simplified.

(a) In general, the catalyst is added in very small amount compared with the reactants.

$S_0 \gg C_0$ so that $S_0 - X \# S_0$

$$K = \frac{|X| |Y|}{[C_0 - X] |S_0|} \quad \text{whence} \quad X = \frac{K |C_0| |S_0|}{K |S_0| + |Y|}$$

$$\mathscr{V} = -\frac{d |S|}{dt} = k_2 |X| |W| = \frac{k_2 K |C_0| |S_0| |W|}{K |S_0| + |Y|}$$

$$= \frac{k_1 k_2 |C_0| |S_0| |W|}{k_1 |S_0| + k_{-1} |Y|}$$

The rate is then always proportional to the concentration of catalyst. In acid-base catalysis, the term in the denominator $k_1 |S_0|$ is always negligible beside $k_{-1} |Y|$, so that the rate equally is proportional to the first power of the concentration of reactants.

(b) In strongly acid or strongly alkaline catalysis, the concentration of catalyst is very much higher than that of reactants $C_0 \gg S_0$ so that

$$K \to \frac{|X| |Y|}{C_0 [S_0 - X]}$$

and

$$\mathscr{V} = -\frac{d |S|}{dt} = \frac{K |C_0| + Y}{k_2 K |C_0| |S_0| |W|}.$$

The rate is always proportional to the concentration of reactant, but tends to become independent of the concentration of catalyst as the latter increases. This appears to have been verified for the acid hydrolysis of acetamide (Euler).

B. *Stationary State*

If the rate of 2 is great, the intermediate X cannot accumulate and the stationary state principle applies

$$\frac{dX}{dt} = 0 = k_1 |C| |S| - k_{-1} |X| |Y| - k_2 |X| |W| = 0$$

so that

$$k_1 |C_0 - X| |S_0 - X| - k_{-1} |X| |Y| - k_2 |X| |W| = 0.$$

If we neglect the infinitely small second order quantity, $|X|^2$, it follows

$$|X| = \frac{k_1 C_0 S_0}{k_1 |C_0| + k_1 |S_0| + k_{-1} |Y| + k_2 |W|}$$

and

$$\mathscr{V} = \frac{d |P|}{dt} = k_2 |X| |W|$$

$$\frac{k_1 k_2 |C_0| |S_0| |W|}{k_1 |C_0| + k_1 |S_0| + k_{-1} |Y| + k_2 |W|}.$$

If k_2 is great, the first three terms in the denominater are in general negligible beside $k_2 |W|$, so that the rate varies linearly with the concentrations of reactants and catalyst.

6.2. Homogeneous Gaseous Catalysis

Many foreign substances affect the rates of chemical reactions in the gaseous phase, even when present in very small amounts. It is sometimes easy to show the presence of intermediate compounds.

EXAMPLE 1: *Formation of nitrogen dioxide* NO_2, in the course of the oxidation of sulfur dioxide or carbon monoxide in the presence of NO.

$$NO \ + 1/2O_2 \rightarrow NO_2$$
$$\begin{cases} NO_2 + SO_2 \ \ \rightarrow SO_3 + NO \\ NO_2 + CO \ \ \ \rightarrow CO_2 + NO \end{cases}$$

EXAMPLE 2: *Oxidation of carbon monoxide* in the presence of water vapour. In the absence of water the process is very slow and only takes place on the container walls (heterogeneous mechanism). In the presence of water the reaction takes place in homogeneous phase.

The pyrolysis of organic substances in the gaseous state is often catalyzed by iodine vapour. The chain mechanism of homogeneous decomposition is then replaced by a series of simple bimolecular reactions between the iodine molecule and the organic substance having much lower activation energies.

EXAMPLE 3: *Pyrolysis of acetaldehyde in the presence of iodine*. The eight steps of the chain mechanism are replaced by the following

$$CH_3CHO + I_2 \rightarrow CH_3I + HI + CO$$
$$\underline{CH_3I \quad\quad + HI \rightarrow CH_4 + I_2}$$
$$CH_3CHO \quad\quad \overset{I_2}{\rightarrow} CO \quad + CH_4.$$

The experimental rate of this pyrolysis is of the form

$$\mathscr{V} = k\,|CH_3CHO|\,|I_2|$$

and the activation energy is 32.5 kcal mole^{-1}, instead of 45.5 in the absence of catalyst.

EXAMPLE 4: *Pyrolysis of ethyl ether.*

$$(C_2H_5)_2O + I_2 \rightarrow C_2H_5I + HI \quad + CH_3CHO$$
$$C_2H_5I \quad\quad + IH \rightarrow C_2H_6 \quad + I_2$$
$$CH_3CHO + I_2 \rightarrow CH_3I \quad + CO \quad + HI$$
$$\underline{CH_3I \quad\quad + HI \rightarrow CH_4 \quad + I_2}$$
$$(C_2H_5)_2O \quad\quad \overset{I_2}{\rightarrow} C_2H_6 \quad + CH_4 + CO.$$

EXAMPLE 5: The decomposition of chloral in the presence of iodine at 353–462°C follows approximately the rate equation

$$\mathscr{V} = k\,|I_2|^{1/2}\,|CCl_3\,CHO|$$

which could be due to the following mechanism.

(1) $I_2 \rightleftarrows 2I^{\bullet}$ rapid preequilibrium
(2) $I^{\bullet} + CCl_3CHO \rightarrow CCl_3^{\bullet} + CO + HI$ slow
(3) $CCl_3^{\bullet} + HI \quad\quad \rightarrow CHCl_3 + I^{\bullet}$ fast

Step 2 is rate-determining for the overall reaction

$$\mathscr{V} = \mathscr{V}_2 = k_2\,|I^{\bullet}|\,|CCl_3CHO|$$

but

$$|I^{\bullet}| = K^{1/2}\,|I_2|^{1/2}$$

according to the equilibrium 1, so that

$$\mathscr{V} = k_2 K^{1/2} |I_2|^{1/2} |CCl_3CHO| \quad \text{with} \quad k = k_2 K^{1/2}.$$

In fact, the actual catalyst for the conversion of chloral into chloroform is the free atom of iodine.

6.3. Catalysis in Solution

Homogeneous catalysis has been especially studied for acids and bases in water or in non-aqueous solvents.

EXAMPLE: *Acid catalysis.*

$$CH_3CO_2C_2H_5 \underset{\text{ester}}{} + H_2O + H^+ \rightarrow CH_3COOH + C_2H_5OH$$

$$CH_3CONH_2 \underset{\text{amide}}{} + H_2O + H^+ \rightarrow CH_3COOH + NH_4^+$$

$$N_2CHCOOC_2H_5 \underset{\text{diazoacetic ester}}{} + H_2O + H^+ \rightarrow HOCH_2COOC_2H_5 + N_2.$$

Base catalysis.

$$(CH_3)_2\overset{|}{\underset{OH}{C}}-CH_2COCH_3 + OH^- \rightarrow 2CH_3COCH_3.$$

diacetone alcohol

Acid-base catalysis

$$\text{glucose } \alpha \quad \rightleftarrows \text{ glucose } \beta \qquad \text{mutarotation}$$

$$CH_3COCH_3 \rightleftarrows CH_3\overset{|}{\underset{OH}{C}}=CH_2 \qquad \text{enolisation.}$$

But many other catalytic reactions, such as the decomposition of hydrogen peroxide by alkalis or by halide ions, have been recognised and studied for more than a century.

EXAMPLE:

$$\begin{array}{ll} H_2O_2 + I^- \quad \rightarrow H_2O + IO^- & \text{slow} \\ \underline{IO^- \; + H_2O_2 \rightarrow H_2O + O_2 + I^-} & \text{fast} \\ H_2O_2 \qquad\;\; \xrightarrow{I^-} H_2O + 1/2O_2 & \end{array}$$

having a rate $\mathcal{V} = k|H_2O_2|\,|I^-|$. At a constant concentration of catalyst, H_2O_2 decomposes following the first-order law $|H_2O_2| = ae^{-k't}$ with $k' = k|I^-|$. The catalytic coefficient k is calculated by measuring rates in the presence of varying concentrations of the ion.

Acid-base catalysis. This catalysis is found in a large number of inorganic, organic, and biological reactions.

Many cases of 'specific' catalysis by protons and OH^- ions have been established, such as the hydrolysis of esters.

Then Bronsted was able to show that other acid or basic entities were able to play quantitatively the same role as protons or hydroxide ions: the catalysis is then designated 'general'.

EXAMPLE: The mutarotation of glucose is a case of general acid-base catalysis, with the rate coefficient for the reaction in water having the form

$$k = k_0 + k_{H^+} \underset{\substack{\text{solvated}\\\text{proton}}}{|H_3^+O|} + k_{OH^-}|OH^-|$$
$$+ k_{AH}\underset{\text{acid}}{|AH|} + k_{A^-}\underset{\text{base}}{|A^-|}.$$

A. *Mechanism of Acid-Base Catalysis*

The scheme already given applies to these two cases, under the form:

Acid catalysis:

$$S + CH \rightleftarrows X + C^-$$
$$X + H_2O \rightarrow P + H_3O^+ .$$

Base catalysis:

$$S + C \rightleftarrows X + CH^+$$
$$X + H_2O \rightarrow P + OH^- .$$

The entities Y and W (in the equations above, H_2O) play a big role: since it is essentially a question of the transfer of protons from the catalyst to one of the reactants and vice versa the substance Y is in fact the conjugate particle of the catalyst C. In the same way, the substance W in acid catalysis acts as a basic or amphoteric substance, capable of removing a proton from the complex X, and in basic catalysis capable of adding a proton to X. This role can be held by a molecule of solvent in the two cases, but equally by other species in solution.

The quantitative treatment by the steady-state principle proves most often convenient, which implies that the formation of the complex is rate-determining; in neglecting for this reason the terms $k_1 |C_0| + k_1 |S_0|$, the rate equation has the form (Section 2).

$$\mathscr{V} = -\frac{d|S|}{dt} = \frac{k_1 k_2 |C_0| |S_0| |W|}{k_{-1} |Y| + k_2 |W|} .$$

Remarks: (1) If $k_2W \gg k_{-1}|Y|$ the formation of the complex controls the process completely and

$$\mathscr{V} \to k_1 |C_0| |S_0|.$$

(2) If $k_{-1}|Y| \gg k_2W$

$$\mathscr{V} \to \frac{k_1 k_2}{k_{-1}} \frac{|C_0| |S_0| |W|}{|Y|}$$

But all the acid and basic entities present in the solution are in equilibrium, with the general equation

$$C + W \rightleftarrows Y + Z \quad \text{with} \quad K' = \frac{|Y| |Z|}{|C| |W|}$$

so that if $|C_0|$ is very little different from $|C|$, in this particular case

$$\frac{|C_0| |W|}{|Y|} = \frac{|Z'|}{K'} \quad \text{and} \quad \mathscr{V} \to \frac{k_1 k_2}{k_{-1}} \frac{|Z| |S_0|}{K'}.$$

If Z is the conjugate acid of the molecule of solvent (e.g. H_3O^+) (or, on the other hand, of another acid solute), it is a question of specific catalysis by protons, or of general acid catalysis, which allows us to class these catalytic reactions equally

as *protolytic* reactions → W = solvent, Z = conjugate acid of solvent, and

as *prototropic* reactions → W = acid solute, Z = conjugate acid of solute.

B. *Superacid and Superbasic Catalysis*

As with reactions in concentrated solutions of electrolytes (Chapter 5) the rate of a catalytic reaction increases exponentially when the concentration of catalyst exceeds about 5×10^{-2} M. The phenomenon is particularly marked with strong acids and basis, which become extremely active in concentrated solution.

It then becomes necessary to take account of the thermodynamic activity of the catalyst, which can be deduced from vapour-pressure measurements above the solution or, for protons or OH^- ions, from the values of the Hammett acidity functions.

The acidity function H_0 of sulfuric acid in aqueous solution, defined by the relation

$$H_0 = - \log \left[a_{H^+} \frac{\gamma_B}{\gamma_{BH^+}} \right]$$

reaches a value of -10 in the pure acid while it becomes identical with pH in dilute solution. It thus allows one to study specific catalysis over a very wide range of rates. For a certain number of reactions, such as the lactonization of γ-hydroxybutyric acid, the rate coefficient follows the relationship

$$\log k_{app} = \log k_0 - H_0.$$

$$CH_2 \begin{array}{c} CH_2{-}COOH \\ \\ CH_2{-}OH \end{array} \rightarrow CH_2 \begin{array}{c} CH_2{-}CO \\ \\ CH_2 \end{array} O + H_2O.$$

Many studies have been made in the last few years in superacid or superbasic solution, because one has available non-aqueous media having acidities (expressed as acidity function) varying between about -13 and $+30$.

6.4. Examples of Catalysis in Solution

A. *Decomposition of Nitramide*

This reaction allowed Bronsted to study general base catalysis. The reaction $NH_2NO_2 \rightarrow N_2O + H_2O$ follows the first-order law, and can be followed by the evolution of nitrous oxide. Insens'ti e to acids, it is on the other hand strongly influenced by bases such as OH^-, CH_3COO^- etc. The mechanism should be the following

$$\begin{cases} NH_2NO_2 + OH^- \rightarrow NHNO_2^- + H_2O \\ NHNO_2^- \rightarrow N_2O + OH^-. \end{cases}$$

B. *Halogenation of Ketones*

The reaction between acetone and iodine was one of the first reactions for which general acid catalysis was established.

The rate of the reaction is independent of the concentration of iodine and depends only on the concentration of ketone and catalyst; it remains the same if iodine is replaced by bromine.

Mechanism:

$$HA + CH_3COCH_3 \underset{k_{-1}}{\overset{k_1}{\rightleftarrows}} CH_3\underset{\underset{OH^+}{\|}}{C}CH_3 + A^-$$

$$CH_3\underset{\underset{OH^+}{\|}}{C}{-}CH_3 + \underset{\underset{solute}{basic}}{B} \to CH_2 = \underset{\underset{\underset{enol}{OH}}{|}}{C}{-}CH_3 + BH^+$$

and the rate of formation of enol is found, by applying the steady-state principle, to be

$$\frac{d\,|enol|}{dt} = \frac{k_1 k_2\,|HA|\,|ketone|\,|B|}{k_{-1}\,|A| + k_2\,|B|}$$

This rate is the overall rate of the halogenation process, as well as of the racemization process (if one starts with an optically active ketone) or of the deuterium exchange reaction. The reaction is prototropic and the rate-determining step is the formation of enol.

$$CH_3COCH_3 + B \to (CH_2COCH_3)^- + BH^+$$

Remark: In the presence of base, another catalytic process occurs by way of a carbanion instead of an enol The carbanion reacts directly with halogen.

C. *Friedel-Crafts Reaction*

This reaction is catalyzed by Lewis acids such as aluminium chloride, and the mechanism assumes the ionic intermediate $R{-}CO^+$.

EXAMPLE:

$$CH_3COCl + AlCl_3 \to CH_3CO^+ + AlCl_4^-$$

$$CH_3CO^+ + C_6H_6 \rightarrow C_6H_5COCH_3 + H^+$$
$$H^+ \qquad + AlCl^- \rightarrow HCl + AlCl_3.$$

Many other examples of catalysis by non-protonic acids are also found

6.5. Catalytic Power and Acidity Constant: Bronsted Relationships

Many attempts were made to connect the catalytic power of acid or basic entities with their thermodynamic dissociation constants. The most satisfactory relationships were proposed by Bronsted

$$k_a = G_a K_a^\alpha \quad \text{(acid catalysis)} \quad \left.\begin{cases} G_a \text{ and } \alpha \\ G_B \text{ and } \beta \end{cases}\right\} = \text{const.}$$

or

$$k_b = G_B' \left(\frac{1}{K_a}\right)^\beta \quad \text{(base catalysis)} \quad 0 < \alpha, \beta < 1$$

where K_a is the ionization constant of the acid-basic catalytic couple. In fact, experiments showed that G_a, G_b', α and β were constants for a given type of catalyst (for example anionic bases) but could have different values for another type (e.g. neutral bases). With this restriction, the formulae allowed the catalytic power of acids and bases to be predicted with adequate precision.

6.6. Autocatalytic Reactions

A final product may have a very noticeable effect on the

course of a reaction. If it accelerates the reaction, it autocatalyzes it; thus, th: hydrolysis of an ester

$$RCOOR' + H_2O \rightarrow RCOOH + R'OH$$

in dilute solution in water goes very slowly at ordinary temperatures and follows a second-order rate law

$$\mathscr{V}_1 = k_1 |H_2O| \, |ester|,$$

as long as the reverse reaction of esterification remains negligible (10 to 20% extent of reaction).

Because the solvent water is in great excess, the reaction in fact follows the first-order law

$$\frac{dx}{dt} = k_1 |ester|$$

but the acid produced during hydrolysis (or more exactly, the proton provided by its dissociation) is a good catalyst for the process, so that a second catalytic reaction develops in parallel with the homogeneous process, having a rate

$$\mathscr{V}_2 = k_2 |RCOOH| \, |ester|.$$

The real process proceeds with an overall rate

$$\mathscr{V} = \mathscr{V}_1 + \mathscr{V}_2 = k_1 |a - x| + k_2 |a - x| \, x.$$

Making $f = x/a$, we obtain

$$\mathscr{V} = (k_1 + k_2 x)(a - x) = (k_1 + k_2 a f)(1 - f)$$

where the inverse of the term $k_2 a$ (product of a bimolecular constant and the initial concentration of ester) is the average time of the catalytic reaction while the inverse of k_1 represents the average time of the homogeneous reaction.

Since $\rho = k_1/k_2 a$ is very small (< 0.01), we obtain

$$\frac{df}{dt} = k(f + \rho)(1 - f) \quad \text{(Ostwald)}.$$

Integrating, taking account of the initial conditions $f = 0$ for $t = 0$, gives

$$f = \rho \, \frac{e^{(1+\rho)kt} - 1}{\rho \, e^{(1+\rho)kt} + 1}.$$

The term $(1 + \rho) \, kt = \mathscr{C}$ is dimensionless and

$$f = \rho \, \frac{1 - e^{(-\mathscr{C})}}{\rho + e^{(-\mathscr{C})}}$$

The curve $f = \text{function } (\mathscr{C})$ (Figure 21) shows an auto-acceleration as far as the inflexion point having the co-ordinates $f = \frac{1}{2}$, $\mathscr{C} = -\log \rho$, which corresponds to a maximum rate, then $f \to 1$ as $\mathscr{C} \to \infty$.

The curve exhibits a characteristic S shape, very different from that of a simple process which slows down constantly with time, and resembles that for the appearance of C by a series of successive reactions $A \to B \to C$ with the accumulation of B; as with the latter, the rate is very low at the start of the process (induction period).

6.7. Inhibition

The term inhibition, a synonym for negative catalysis,

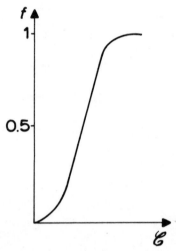

Fig. 21.   Autocatalytic reaction.

corresponds to a slowing of reaction processes by foreign substances or by final reaction products.

The cases known in homogeneous kinetics correspond to radical processes, and inhibitors function as traps of active centres, thus preventing the propagation of the chain.

EXAMPLE 1: Inhibition by hydriodic acid of reactions proceeding by hydrogen atoms.

$$IH + H^{\cdot} \rightarrow H_2 + I^{\cdot}.$$

The iodine atom is unreactive, and preferentially combines with itself

$$2I^{\cdot} + M \rightarrow I_2 + M.$$

EXAMPLE 2: The oxidation of sodium sulfite or benz-aldehyde is retarded by the addition of small amounts of alcohols such as mannitol or benzyl alcohol.

These inhibitors have very great practical importance, because they allow many products to be stabilized (e.g. antioxidants in the rubber industry). Radical polymerisations can be inhibited by various substances; often the presence of impurities causes a longer or shorter induction period (in which polymerisation is halted), until the impurities have been entirely consumed by reaction with free radicals.

Inhibition by reaction products is relatively common in closed sequences, and has been described in the synthesis of hydrobromic acid, where inhibition by HBr was treated quantitatively.

6.8. Radial Processes and Catalysis

When the reaction process in closed sequence accelerates in the presence of a catalyst, the change in rate is no longer linear with respect to concentration of the latter.

Two cases can be envisaged:

(a) The thermal initiation reaction is replaced by a bimolecular reaction between the molecule M of reactant and the catalyst.

The rate of this catalytic initiation has then the usual form

$$\mathscr{V}_i = k_i |C| \, |M|$$

but the overall rate of the chain process has then the form

$$\mathscr{V} = k |M|^{3/2} |C|^{1/2}$$

and varies as the square root of the concentration of catalyst (see following examples)

(b) The rate of catalytic initiation can be of the form

$$\mathscr{V}_i = k_i |C|$$

(first-order with respect to catalyst, zero order with respect to reactant) which leads to the following expression for the overall rate

$$\mathscr{V} = k |M| \, |C|^{1/2}$$

In the two cases the rate is proportional to the square root of the concentration of catalyst.

EXAMPLE. *Decomposition of ozone in the presence of chlorine.*

The process:

Initiation $O_3 + Cl_2 \rightarrow ClO + ClO_2$
Propagation $ClO_2 + O_3 \rightarrow ClO_3 + O_2$
                    $ClO_3 + O_3 \rightarrow ClO_2 + 2O_2$
Termination $ClO_3 + ClO_3 \rightarrow Cl_2 + 3O_2$

involves chain carriers ClO_2 and ClO_3 whose concentrations can be calculated by the steady state principle

$$|ClO_3| = \left[\frac{k_1 |Cl_2| |O_3|}{k_4}\right]^{1/2}$$

$$|ClO_2| = \frac{k_3}{k_2}\left[\frac{k_1 |Cl_2| |O_3|}{k_4}\right]^{1/2} + \frac{k_1}{k_2} |Cl_2|$$

which gives for the overall rate

$$-\frac{d|O_3|}{dt} = 2k_3 \left(\frac{k_1}{k_4}\right)^{1/2} |Cl_2|^{1/2} |O_3|^{3/2} + 2k_1 |Cl_2| |O_3|$$

which agrees with the experimental rate, $k|O_3|^{3/2}|Cl_2|^{1/2}$, if one neglects the last term (as is legitimate if the chain is sufficiently long).

EXAMPLE. *Polymerization reactions.*

Polymerizations take place by a chain process starting from an unsaturated monomer. The initiation reaction gives a free radical or an ion which adds to the double bond of the monomer to give a second radical or ion

$$R^{\cdot} + X{-}CH = CH_2 \rightarrow X{-}\overset{\cdot}{CH}{-}CH_2{-}R$$

$$X{-}\overset{\cdot}{CH}{-}CH_2R + XCH = CH_2$$
$$\rightarrow X{-}\overset{\cdot}{CH}{-}CH_2{-}\underset{\underset{X}{|}}{CH}{-}CH_2R$$

The chain is propagated in this way, until interrupted by recombination of radicals, according to the scheme

$$R_1^\cdot + M \rightarrow R_2^\cdot$$
$$R_2^\cdot + M \rightarrow R_3^\cdot$$
$$\cdot \quad \cdot \quad \cdot \quad \cdot \quad \cdot \quad \cdot \quad \cdot \quad \cdot \quad \cdot$$
$$R_m^\cdot + R_n^\cdot \rightarrow M_{m+n}$$

and the molar mass of radicals $R_1^\cdot R_2^\cdot R_3^\cdot$ increases up to a limit set by termination reactions.

Kinetically, the steady-state principle is applicable to each radical, and gives the following, if we call the rate of the initiation reaction $\mathscr{V}_i$, and make the plausible hypothesis that all termination reactions have the same rate coefficients k_t and all propagation reactions the same coefficients k_p

$$\mathscr{V}_i - k_p |R_1| \, |M| - k_t |R_1| \, \Sigma |R_n| = 0$$
$$k_p |R_1| \, |M| - k_p |R_2| \, |M| - k_t |R_2| \, \Sigma |R_n| = 0$$
$$k_p |R_2| \, |M| - k_p |R_3| \, |M| - k_t |R_3| \, \Sigma |R_n| = 0.$$

After adding up all the equations above, there remains

$$\mathscr{V}_i - k \, \Sigma |R_n| \, \Sigma |R_n| = 0$$

so that

$$\Sigma |R_n^\cdot| = \sqrt{\frac{\mathscr{V}_i}{k_t}}$$

and

$$\mathscr{V}_{\text{overall}} = -\frac{d|M|}{dt} = k_p |M| \, \Sigma |R_n| = k_p \sqrt{\frac{\mathscr{V}_i}{k_t}} \, |M|$$

– If initiation is purely thermal

$$\mathscr{V}_i = k_i |M|^2 \quad \text{and} \quad \mathscr{V} = k_p \sqrt{\frac{k_i}{k_t}} \, |M|^2$$

(case of styrene in solution and in the gaseous phase);
– If initiation is catalytic, of the form $M + C \rightarrow R^{\bullet} + prod$-
ucts

$$\mathscr{V}_i = k_i |M| \, |C| \quad \text{and} \quad \mathscr{V} = k_p \sqrt{\frac{k_i}{k_t}} \, |M|^{3/2} \, |C|^{1/2}$$

(case of styrene and vinyl acetate in the presence of benzoyl peroxide, of ethylene in the presence of azomethane);
– If initiation is catalytic, of the form $C \rightarrow R^{\bullet} + products$

$$\mathscr{V}_i = k_i |C| \quad \text{and} \quad \mathscr{V} = k_p \sqrt{\frac{k_i}{k_t}} \, |M| \, |C|^{1/2}$$

(case of butyl α-chloroacrylate).

The catalysts of radical processes must be able to decompose easily to give free radicals; examples are chlorine, azomethane, benzoyl peroxide. While thermal initiation reactions are very slow $(E \# 40 \text{ kcal mole}^{-1})$ propagation reactions are very fast $(1–3 \text{ kcal mole}^{-1})$, and it is advantageous in industrial practice to operate with catalysts such as peroxides.

Furthermore, the length of the chain which determines the degree of polymerization depends on the concentration of catalyst

$$l = \frac{k_p}{\sqrt{k_t \mathscr{V}_i}} \, |M| = \frac{k_p}{\sqrt{k_t k_i}} \sqrt{\frac{M}{C}} \, .$$

6.9. Catalysis and Ionic Polymerization

When polymerization reactions take place by heterolytic processes, they show the usual characteristics of such reactions in solution (influence of dielectric constant, etc.), and can be influenced by ionic or strongly polar catalysts.

The cationic polymerization of isobutene, styrene, and alkyl vinyl ethers is strongly influenced by Lewis acids such as $AlCl_3$, BF_3, $TiCl_4$, $ZnCl_2$.

EXAMPLE:

$$ROH + BF_3 \rightarrow F_3B \rightarrow O{\Big\langle}^{H}_{R} \rightleftarrows F_3B \rightarrow \bar{O}R + H^+$$

$$H^+ + XCH = CH_2 \rightarrow X^+CH - CH_3 \text{ (carbonium ion)}$$

which in turn adds to the monomer. This type of catalysis is an example of acid catalysis by proton transfer involving a co-catalyst ROH.

Anionic polymerizations of methyl methacrylate and of methylacrylonitrile are due to carbanions according to the scheme

$$\left(\begin{array}{c} X \\ | \\ RCH_2-C \\ | \\ Y \end{array}\right)^{-} + CH_2 = C\,|X|\,Y$$

$$\rightarrow \left(\begin{array}{cc} X & X \\ | & | \\ RCH_2-C-CH_2-C \\ | & | \\ Y & Y \end{array}\right)^{-}$$

The carbonions form through the action of catalysts such as $RMgX$, RNa, KNH_2, alkali metals in solution in ether or amines, etc.

BIBLIOGRAPHY

Amis, S. E., *Solvent Effects on Reaction Rates and Mechanisms*, Academic Press, New York, 1966.

Amdur, I. and Hammes, G. G., *Chemical Kinetics*, McGraw Hill, New York, 1966.

Ausloos, P. J., 'Ion Molecule Reactions in the Gas Phase', *Am. Chem. Soc. (Adv. Chem. Series 58)*, 1966.

Bamford, C. H., *Chemical Kinetics, I and II*, Elsevier, Amsterdam, 1969.

Boudart, M., *Kinetics of Chemical Processes*, Prentice Hall, New York, 1968.

Chance, B., *Rapid Mixing and Sampling Techniques in Biochemistry*, Academic Press, New York, 1964.

Jungers, J. C., *Cinétique chimique appliquée. Analyse cinétique de la transformation chimique*, Technip, Paris, 1959, 1967.

Laidler, K. J., *Chemical Kinetics*, McGraw Hill, New York, 1950.

Pannetier, G. and Souchay, P., *Cinétique chimique*, Masson, Paris, 1964.

Rodiguin, N. M. and Rodiguina, E. N., *Consecutive Chemical Reactions*, Van Nostrand, Princeton, 1964.

Weissberger, A., *Investigation of Rates and Mechanisms of Reactions, I and II*, Interscience, New York, 1963.